Design and Construction of Bioclimatic Wooden Greenhouses 2

Series Editor
Gilles Pijaudier-Cabot

Design and Construction of Bioclimatic Wooden Greenhouses 2

Design of Construction: Structure and Systems

Gian Luca Brunetti

WILEY

First published 2022 in Great Britain and the United States by ISTE Ltd and John Wiley & Sons, Inc.

Apart from any fair dealing for the purposes of research or private study, or criticism or review, as permitted under the Copyright, Designs and Patents Act 1988, this publication may only be reproduced, stored or transmitted, in any form or by any means, with the prior permission in writing of the publishers, or in the case of reprographic reproduction in accordance with the terms and licenses issued by the CLA. Enquiries concerning reproduction outside these terms should be sent to the publishers at the undermentioned address:

ISTE Ltd
27-37 St George's Road
London SW19 4EU
UK

www.iste.co.uk

John Wiley & Sons, Inc.
111 River Street
Hoboken, NJ 07030
USA

www.wiley.com

© ISTE Ltd 2022

The rights of Gian Luca Brunetti to be identified as the author of this work have been asserted by him in accordance with the Copyright, Designs and Patents Act 1988.

Any opinions, findings, and conclusions or recommendations expressed in this material are those of the author(s), contributor(s) or editor(s) and do not necessarily reflect the views of ISTE Group.

Library of Congress Control Number: 2022941486

British Library Cataloguing-in-Publication Data
A CIP record for this book is available from the British Library
ISBN 978-1-78630-852-8

Contents

Introduction ix

Chapter 1. Light Frames (Wooden Frames) 1

    1.1. Commonest solution: platform-frame-like or balloon-frame-like curtain walls framed with studs/mullions 6

    1.2. Types of connections in wooden construction 7

        1.2.1. Head-to-head butt joint 7

        1.2.2. Head-to-side butt joint 9

        1.2.3. Lap joints 11

        1.2.4. Nailed connections 12

        1.2.5. Screwed connections 14

        1.2.6. Bolted connections 14

        1.2.7. Tooth plate connections 14

        1.2.8. Glued connections 15

    1.3. Types of connections between structural sub-systems 15

        1.3.1. Interfacing the frames of the front façade and the roof 15

        1.3.2. Interfacing the frames of the side walls and the roof. 17

        1.3.3. Gable walls 21

        1.3.4. Opaque enclosures 29

        1.3.5. Back walls 30

        1.3.6. Connection between the roof of a building and the roof of the greenhouse attached to it 42

    1.4. Alternative structural solution: light-frame curtain walls supported by horizontal purlins 42

    1.5. Alternative structural solution: trussed light-frame structures 43

        1.5.1. Trusses and trussed portals or semi-portals orthogonal to the front façade 44

        1.5.2. Light-frame trussed portals parallel to the front façade 44

1.6. Criteria for the construction of light-frame trusses and trussed portal frames . 45
    1.6.1. Chords, diagonals and uprights overlapped on different planes. 45
    1.6.2. Chords, diagonals and uprights on the same plane. 45
    1.6.3. Transversal connection of portal frames 46
    1.6.4. Bracing strategies in light frames 46
1.7. Intermixing parts of timber frames into light frames 51
1.8. Analogies with cold-rolled light frames 52
1.9. Arched and vaulted construction in light frames 53
    1.9.1. Lamella vaults. 53
    1.9.2. Geodesic domes. 57

Chapter 2. Timber Frames. 61

2.1. Intermixing light-frame parts into timber frames 62
    2.1.1. Light frame completely additional to the timber frame 63
    2.1.2. Combining timber frames and light frames. 64
2.2. Connections in timber-frame greenhouses. 75
    2.2.1. Traditional connections in timber frames 76
    2.2.2. Modern connections in timber frames. 77
2.3. Structural solutions with the primary beams of the frames orthogonal
to the front façade. 110
    2.3.1. Post-and-beam greenhouses with primary beams perpendicular
    to the front façade 111
    2.3.2. Trussed post-and-beam greenhouses with primary beams perpendicular
    to the front façade 118
    2.3.3. Portal frames perpendicular to the front façade 120
    2.3.4. Spans of the secondary structural elements in greenhouses having the
    principal beams orthogonal to the main façades. 129
    2.3.5. Frames or portal frames, solid or trussed, parallel to the front façade . . . 130
2.4. Pole construction. 132
    2.4.1. Treating timber poles for a longer life span. 132
    2.4.2. Solutions for cantilevering the poles from the ground 134
    2.4.3. Solutions for connecting girders or beams to the poles 136
    2.4.4. Pole greenhouse construction 137
2.5. Bracing strategies in timber frames 149
    2.5.1. Bracing with cables or rods 151
    2.5.2. Bracing with short massive diagonals. 152
    2.5.3. Bracing the bays with full-length diagonals connected with butt joints
    by means of steel plates 152
    2.5.4. Bracing with full-length lap-joined diagonals 152

Chapter 3. Foundations 161

3.1. Foundation walls and foundation sills 161
3.2. Construction strategies for foundation walls 168
    3.2.1. Preparing the ground for a foundation wall. 169
    3.2.2. Boulders-and-mortar wall foundation. 169
    3.2.3. Brick masonry wall foundation 170
    3.2.4. Conventionally mortared hollow concrete block wall foundation 170
    3.2.5. Parged hollow concrete block wall foundations 172
    3.2.6. Concrete foundation walls. 172
    3.2.7. Wooden-frame foundations 172
    3.2.8. Timber foundations. 175
    3.2.9. Pier foundations. 184
    3.2.10. Insulation of the foundation wall. 184
    3.2.11. The foundation wall as a sill 186
3.3. Drainage around the foundation wall 186
3.4. Pavements 187
3.5. Platform frame floors raised above the ground 188

Chapter 4. Heating and Cooling Systems; Watering Systems 191

4.1. Heating and cooling plants 191
    4.1.1. Air-based systems. 191
    4.1.2. Water-based systems 193
4.2. Heat recovery via air-to-air heat exchangers 193
4.3. Passive and low-energy heating and cooling solutions based on the thermal exchange with the ground 194
    4.3.1. GAHT systems 195
    4.3.2. Ground-air heat exchangers – Canadian wells 197
    4.3.3. Considerations about the transfer of heat to remote masses by convection 199
    4.3.4. Surface air-to-ground heat exchange (experimental) 202
4.4. Auxiliary heating systems. 204
    4.4.1. Electric heating 206
    4.4.2. Common stoves. 209
    4.4.3. Rocket mass stoves 209
    4.4.4. Water systems coupled with burners or heat pumps 209
    4.4.5. Active systems using renewable energy sources 210
    4.4.6. Heat pumps 211
4.5. Auxiliary cooling systems 211
4.6. Integration of photovoltaic panels in greenhouses 213
4.7. Integration of passive solar heating panels in greenhouses. 214

4.8. Watering systems 215
    4.8.1. Most common water sources. 216
    4.8.2. Water containers 217
    4.8.3. Water distribution. 217
4.9. Solutions for water catchment and storage suitable for self-building 219
    4.9.1. Creation of low-cost ponds. 220
    4.9.2. Rainwater collection 221

Conclusion 225

References. 227

Index. 241

Summaries of other volumes. 245

Introduction

The first volume of this work laid out the basis for bioclimatic greenhouse design; and the present one – the second in the series – deals with the construction design of greenhouse timber structures, covering a range of issues spanning from frame configurations, connections, the consequences of precision (or the lack of it), to auxiliary mechanical heating and cooling systems. The contents of this volume will add concreteness to the knowledge basis laid out in the first volume, and also widen operational opportunities. This is because usually design ideas can be addressed in more than one way, each of which can reverberate consequences back to primary design decisions. From this perspective, preliminary design ideas and detailed constructional ideas can be seen as the two sides of the same coin. These two sides will be examined separately in this work for the sake of the analysis, but their distinction tends to be very blurred in action.

Another characteristic of this volume is that it devotes special attention to self-buildable solutions. This is done with two aims. The first is to address not only the professional reader, but also the non-professional reader; and the second aim is to demystify construction techniques. This has been done considering that, due to the pragmatic nature of constructional knowledge, constructional knowledge tends most often to be passed on to others through narrations that only very loosely relate causes and effects (constructional prescriptions). By emphasizing the aspects of self-buildability of greenhouse frames, this volume looks for the causal relations of things and attempts to demystify the stiffness and unavoidability of construction "rules" and rules-of-thumb – modern as well as traditional.

The most important thing to consider about the constructability of greenhouse structures is that the connections between wooden elements constituting the frames (beams, columns, studs, rafters, purlins) – at least, the connections which are not arranged in trusses – are not resistant to bending: that is, they can only – or mainly –

transmit shear[1]. To transmit bending forces, timber components should usually be made rigid by arranging diagonal elements in triangular shapes. This may seem to be a strong limitation, but it is not; indeed, this does not prevent the possibility of deriving many constructional solutions from the primary considerations, and potentially evolving them beyond recognition.

One final note: as with all the other volumes composing this work, this volume will also deliver, together with a basis for sedimented knowledge, some experimental, innovative contents. The reader will be made aware in due places of the experimentality of the contents, so as to make them aware of the necessity of taking extra care in handling them. In this second volume, key examples of such experimentality are constituted by the trussed configurations presented in Chapter 2, and by the timber log and gravel trench foundation presented in section 3.2.8.1.

On references

The book by Alward and Shapiro (1980), despite being more than 40 years old and mostly featuring single-sheet enclosures, due to the level of interest and variety of the solutions it proposes and the constructional "taste" involved, is still a great reference as regards the constructional design of timber solar greenhouses of the lean-to type. The book by Shapiro (1984) is closely related to that by Alward and Shapiro (1980), but is wider in scope, because it not only covers the domain of construction, but also involves a multidisciplinary approach to design. The book by Neal (1975) is an antecedent of these works.

Other essential references for the present work are the book by Marshall (2006), which is a thorough guide to platform-frame style greenhouses; the book by Fischer and Yanda (1977), which focuses on low-cost solutions for platform-frame attached wooden greenhouses, experimented diffusely at the beginning of the 1970s; the book by Herzog and Natterer (1984 – in French and German), which develops a highly cultured architectural analysis on the integration of attached wooden greenhouses and buildings. The volume of proceedings edited by Hayes and Gillet (1977) conveys the breadth and depth of the debate about solar greenhouses in the second part of the 1970s.

Other references that influenced the writing of this book are as follows.

The structural and functional characteristics of European climate greenhouses were investigated in review articles by von Elsner et al. (2000a, 2000b). Lagier and Dastot (2008) is a lean reference of techno-scientific advice.

---

[1] Because of this, in Volume 4, section 2.4, when calculating the wooden structures, we will assume that the bending moments at the connections are null or 0.

The book by Haupt and Wiktorin (1996 – in German) analyzes the constructional side of greenhouse design, making reference primarily to steel structures, and also touching upon wooden structures. Haupt (2001 – in German) keeps a similar line of analysis, but with a briefer theoretical treatment and extended range of examples based on built case studies. Zappone (2009 – in Italian) focuses especially on the thermal design of lean-to greenhouses, also including construction considerations.

The book by Schwolsky and Williams (1982) explores the construction side of passive building design, mostly featuring timber and wooden frames, while the book by Temple and Adams (1981) focuses more specifically on solar collection systems.

The books by Lorenz-Ladener (2013) and Schiller and Plincke (2016) are comprehensive guides to the design of solar greenhouses, also taking into account constructional aspects. The books by Drexel (1999 – in German), Fisch (2001), Stempel (2008 – in German) and Price and Greer (2009) are references delivering pure constructional advice.

The books by Jones (1978), National Research Council of Canada (1981), Mauldin (1987), Kolb (1990 – in German), Lees and Heyn (1991), Schmidt (2011) and Toht (2013) are technical references bringing novelties and inventiveness about construction solutions into the general scenery.

The books by Kurth and Kurth (1982), Freeman (1994, 1997) and Nengelken (1996 – in German) touch upon construction considerations, but are focused on the matter of greenhouse gardening and growing.

The books by Williams (1980), Wolpert (1989 – in German), Bastian (2000 – in German) and AAVV (2002) keep a middle ground between greenhouse gardening and architectural construction, while Jeni (2005 – in German) and Timm (2000 – in German) lean completely towards architecture.

The booklet by Matana (1999 – in French) regards attached greenhouses, and is closely related to the aspects of the French market. The booklet by Ibbotson (1964) is a constructional reference (today, historical) featuring the first outcomes of experimentation with polyethylene films.

The books by IEA (1999, 2000) lay out scientifically referenced criteria and examples for designing and constructing passive solar collector systems of all types, including greenhouses. The books by Kornher and Zaugg (2006) and Smith (2011) are useful guides to the design and construction of passive solar collectors. The publication by the New York State Energy Research and Development Authority (1982) features three highly detailed projects for the construction of solar air collectors to be hung on walls.

Carter (1981) provides design and construction advice for solar building renovation, with an emphasis on problems regarding residential wooden buildings. Wing (1990) takes a similar approach, with an emphasis on extension and remodeling, but without specific passive solar implications.

Kern (1975) is a seminal book about alternative construction technologies, and Twitchell (1985) collects interesting cases of solar architecture from the solar "golden age" of the 1970s.

Pracht (1984 – in German, and translated into French) is a sophisticated, technically complete and tasteful book on wooden construction. Herzog et al. (2004 – in German, and translated into English) is a book on wood construction complete with examples. Kolb (2008 – in German, also translated into French) is a complete reference on the technical evolution that occurred after the turn of the millennium. Moro (2009 – in German) is filled with precious information about timber construction and façade design. Wagner and DeKorne (2002) and Miller (2004) are sources of guidance for self-construction filled with practical advice. Along the same line, Carrol (2005) focuses on techniques and tips for solo building.

The books by Oesterle et al. (2001) and Bonham (2019) provide design and construction advice on double skins in high-rise buildings, and Compagno (2002) provides design and construction advice about the automated management of high-tech building façades.

1
Light Frames (Wooden Frames)

A light frame is a frame constituted by diffusely distributed elements held together by many connections, usually simple to execute and subject only to small forces. Some building norms for greenhouses (like the current ones in use in the European Union)[1] distinguish structures subject to small displacements from structures subject to large ones, and greenhouse frames in general (both light frames and timber frames – i.e. heavier frames) fall into the first category – that of the structures subject to small displacements. Post-and-cable structures, hoop-and-cable structures, tent structures and any case of structures involving cables as primary components can, instead, fall into both categories, being often allowed to have considerable displacements under wind loads.

Light-frame greenhouses are greenhouses in which the vertical elements that bear the transparent enclosures (glass panels, or plastic panels, of plastic films) also support the greenhouse structure. Vertical supports in light frames are indeed directly constituted by studs or mullions rather than by columns. Light frames are therefore the simplest of construction solutions for a greenhouse. This simplicity, however, comes at the price of requiring the designer to strike a compromise between aims related to different functions: in particular, the load-bearing function performed at the greenhouse level, the function of holding in place the elements of the transparent enclosures, and other functions like assuring air-tightness, water-tightness, and durability. As regards the terms "stud" and "mullion" as they are used in this book: the two types of construction elements can be similar and even identical, and which of the two terms is most appropriate depends on the context.

For a color version of all figures in this book, see www.iste.co.uk/brunetti/greenhouses2.zip.

1 EN 13031:2019 – Commercial production greenhouses. Parts 1 and 2.

The term "stud" is commonly used for lean vertical supports that are enclosed between opaque panels, while "mullion" commonly designates the lean vertical supports to which the transparent enclosures are anchored, and that remain in plain sight. Dimensions of typical sections of stud and mullions vary, but timber sections of 5x10, 6x10, 6x12, 5x15, 6x15 or 6x18 cm are fairly common for small greenhouses.

Under many viewpoints, the light-frame construction of a greenhouse resembles that of a balloon-frame house (a light-frame house in which the studs extend from the top of the foundation wall to the roof, even in multi-story configurations). However, the light frames of greenhouses are commonly enclosed with transparent panels one structural bay wide, rather than opaque plywood panels or planks two or more bays wide.

In this context, horizontal wooden plates are positioned at the foot of the studs/mullions (as mediation elements between the studs/mullions and the foundation walls/sill plates), and at the top of them (as "ties" holding them together), and each roof rafter (or roof joist) is supported directly by the studs/mullions beneath it, with the horizontal plate in the intermediate position (see Figure 1.1). However, the alternative also exists of directing the load of the rafters onto the header joist (or fascia beam), or, in any case, onto a header edge beam, rather than directing it onto the plate on top of each stud/mullion. The described solution constitutes an equivalent to the one commonly adopted in platform frame structures, which is also the simplest and cheapest one that is available.

The construction sequence can follow two paths: (a) the studs/mullions can be put in place vertically on the wooden bottom plates, keeping them vertical with bracing elements, then joining them with a top plate; or (b) the stud/mullions can be used for building the wall frames horizontally (on-site or off-site), and then the frames can be tilted up vertically in their definitive position. The latter solution, which is typical of platform frames, is the most common. In both cases, the bottom plates are composed of a lower plate (usually constituted by pressure-treated lumber) anchored with anchor bolts to the foundation wall, and of a top plate (constituted by ordinary construction lumber, and belonging to the wall's frame) nailed or screwed to the studs/mullions. Similarly, the top plates can be composed of a bottom plate (tie beam) belonging to the wall frame, anchored to the studs/mullions, and by a top plate (header plate) holding together the plates below (being lap-joined with it, overlapped with the joints non-coincident), as well as the whole framed wall panels (Figure 1.1), or they may be constituted by a single component.

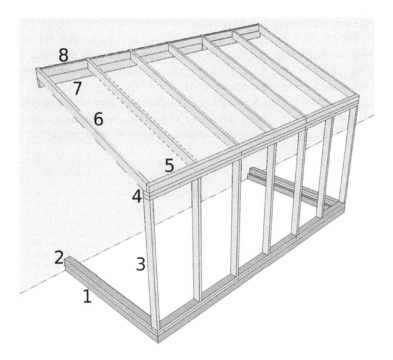

Figure 1.1. *The ordinary lean-to-greenhouse structure is built in a platform-frame-like manner under all aspects. In gray, the wall shared with the building. The structure of the gable wall is not shown. Legend: 1. Strip foundation; 2. Bottom plates (below: treated lumber; above: construction lumber); 3. Mullion; 4. Top plates (two, one on top of the other, with the joints not coincident – the plate below is also known as the "tie beam"); 5. Header plate (or fascia beam); 6. Rafter; 7. Ledger (continuous); when anchored to a masonry wall or a concrete wall, the anchorage can be obtained through expansion dowels; 8. Stop between rafters (non-continuous)*

It is interesting to note that a thorough knowledge of the criteria for the construction of light-frame greenhouses is important not only because light-framed greenhouses are common (especially in the context of small-size greenhouses), but also because light frames are used in timber-frame greenhouses as well (see Chapter 2). Indeed, timber-frame greenhouses often combine timber-frame solutions and light-frame ones. Indeed, as it will be seen in the next chapter, in timber frames, the relation between the structure and the transparent enclosures is usually solved with the aid of secondary structures of the light-frame type – utilizing mullion, transoms, purlins and/or rafters.

The construction of light-frame greenhouses is simpler than that of timber-frame ones, because it requires applying only one construction method; but the sizing of light frames can sometimes be a rather delicate operation, deceptively simple, except when it involves short spans and short heights – in which case, the oversizing of the frame related to the fact that the frame is performing in the same time as a load-bearing structure and as a skeleton of the envelope is unnoticeable. This happens because the framing elements used in light frames have to satisfy both the general sizing requirements derived from the fact that they are greenhouse-scale structures, and the specific, localized sizing and joining requirements regarding the transparent enclosures and their connections, related to their nature of envelope systems.

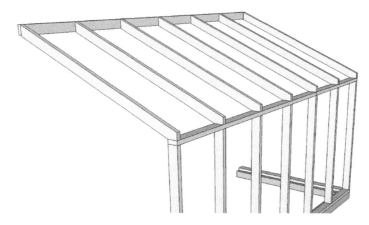

Figure 1.2. *Example of the simplest possible relation between front façade and roof in a light-frame structure of an attached greenhouse. Each rafter rests on a mullion/stud, via a top plate anchored to the foundation (or on ring beams, in arrangements above ground level). How the roof rafters are connected to the back wall (here not shown) differs from the way shown in the previous example: indeed, they are anchored to a ledger at the back of the roof, which is, in turn, anchored to the façade*

Figure 1.3. *In this example, the rafters are anchored to the back wall (here not shown) via a ledger positioned below them, like in the example in Figure 1.1; but here the stops between the heads of the rafters are not used*

Light Frames (Wooden Frames)   5

Figure 1.4. *In this solution, the plate on top of the studs/mullions has been doubled but, unless the two plates in question are connected so as to collaborate in bending (which is not usual, but may be obtained by diffuse nailing or screwing), each rafter still has to rest on the mullion/stud below*

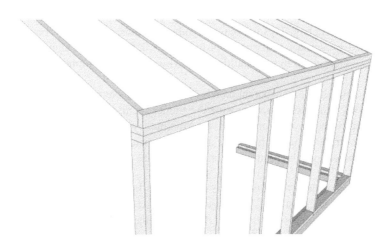

Figure 1.5. *Combining the top plates with a header joist (fascia beam) adds rigidity to the structure at the eave and reduces the need to make each rafter rest on the mullion/stud below*

On references

The book by Allen (2014) is very strong on platform frame techniques and has been a useful reference in writing this book. Burrows (1967–2013), Haun (1998), Wagner (2005), Benoit (2014 – in French) and Thallon (2016) are high-quality books specifically focusing on the platform-frame construction technique. Feirer et al. (1997), Canadian Wood Council (1991), Spence (1999) and Engel (2011) are thorough references about wooden carpentry and joinery, covering both platform-frame and timber-frame construction.

1.1. Commonest solution: platform-frame-like or balloon-frame-like curtain walls framed with studs/mullions

Generally, the spacing of the studs/mullion is defined on the basis of the following criteria: (a) in the case in which glass panels are used as transparent enclosures, the space between mullions should be slightly smaller (at least approximately 1–1.5 cm per side, depending on the quantity of movement that the structure is allowed to undergo) than the size of the glass panels, in order to allow for a sufficient overlap of the glass panels themselves to the frames (Figure 1.9, Volume 3). (b) In the case in which synthetic transparent panels (polycarbonate, acrylic, fiberglass) are used, the space between mullions may vary from slightly less than the width of the transparent panels (about 1–1.5 cm, similarly to what is required in the case of glass panels) to a fraction of it (minus a space of again at least about 1–1.5 cm per side for the overlap of the panels at the side), when the panels are wider than one bay. This is because synthetic panels, unlike glass ones, can also be traversed by screws, nails or bolts in intermediate positions along their width. This allows their width to span multiple bays (Figure 2.11, Volume 3).

In this context, the transparent panels (of whatever material) are usually anchored to the frames by pressing them against the frames themselves, as seen, by means of pressure caps (that may be constituted, for example, by wooden plates, C-shaped steel profiles or extruded aluminum profiles), with the mediation of gaskets on both faces of the transparent panels (Figures 1.2 and 1.3, Volume 3).

In the most ordinary solutions, the connections between the frame elements are performed by means of nails or screws positioned orthogonally to the elements, when possible (Figures 1.10 and 1.14), and otherwise are nailed or screwed diagonally (Figures 1.6, 1.7, 1.8, 1.9, 1.11 and 1.4). Statically speaking, single-nailed or single-screwed connections work like hinges, and multiple ones may not, but in any case, a light-frame connection lacks the rigidity which is necessary to obtain rigid connections unless the configuration of the connections is trussed (i.e. it contains triangles). And constructing frames without rigid connections requires bracing them for stability against horizontal forces. To this, it has to be added that, when trussed configurations are used, because light frames are composed of small-section, low-resistance elements, for bracing them, usually it is not a good idea to rely on short and strong, "concentrated" bracing members spanning only near the zone of the connection (of the kind of schemes in the right-hand half of Figure 2.104). In those situations, the opposite is usually appropriate: diffuse, distributed diagonals involving the whole span of a structural bay, or even spanning several bays (as in the schemes in the left-hand half of Figure 2.104), should be more appropriately used.

1.2. Types of connections in wooden construction

Connections in light frames are conceived to make the construction operations as simple as possible. This is what modern carpentry is often about. They are therefore executed by simple nailing or screwing. Nailing and screwing, however, are also used in timber-frame connections. This shows that the distinction between light-frame connections and a timber-frame one is not precise. Indeed, it is less precise than that between light-frame structures and timber-frame ones. There is a great deal of middle ground between the two types of technique.

The most common light-frame connections (which are also the most common in timber frames) are the butt joint and the lap joint. The butt joint can be of the head-to-head type (Figures 1.6 and 1.8) or of the head-to-side type (Figure 1.10). Both butt joints and lap joints produce non-bending-resistant connections which, statically speaking, can be interpreted as pinned connections. This is not due to the fact that the nails or screws could not be positioned so as to obtain a lever arm large enough to make the connections rigid, bending resistant, but to the fact that the individual connections of which a composite connection is made are usually not strong enough to make the derived connection bending resistant and at the same time simple.

1.2.1. *Head-to-head butt joint*

Head-to-head butt joints are usually not suitable for execution by means of simple nailing or screwing (Figures 1.6–1.9), but rather by means of cover plates: which, in the simplest (and weakest) configurations, are asymmetrical (i.e. placed on one side of the connection only), and in the strongest configurations are symmetrical (i.e. placed on both sides) (Figure 2.28). The cover plates add a certain rigidity to the connections in such a way that the longer and more overlapped they are, the more bending resistant and rigid the connections are (Figure 2.32).

For executing a butt joint without plates, there is no choice other than placing the nails or screws diagonally in the components, so as to position them in directions as mutually divergent as possible, so as to make them offer resistance against extraction (Figures 1.6–1.9). At a statical level, what is obtained is, in any case, again a pinned joint, of a kind, though of course more resistant in compression than in tension.

The cover plates can be made of steel (Figure 2.28) or wood (Figure 2.31).

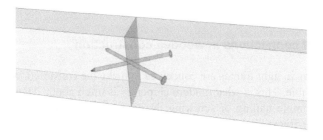

Figure 1.6. *Head-to-head butt joint side-nailed horizontally. A weak, shear-only-resistant connection*

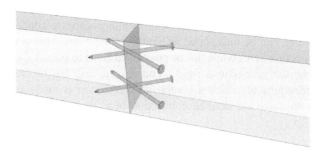

Figure 1.7. *Head-to-head butt joint double-side-nailed horizontally*

Figure 1.8. *Head-to-head butt joint nailed vertically. Another weak, shear-only-resistant connection*

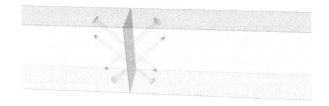

Figure 1.9. *Head-to-head butt joint double-nailed vertically*

1.2.2. *Head-to-side butt joint*

In the case of the head-to-side butt joint, one end of one element is positioned at the side of the other. This connection can be performed again (a) by means of two cover plates (Figure 2.28); (b) by means of one or two angles (single and asymmetrical, in the case of the weakest connection type; double and symmetrical, in the case of the strongest – Figure 2.27); angles which, in the case of light frames, are nailed or screwed, while in the case of timber frames, are bolted or screwed; and (c) by means of only screws or nails. In this case too, the nails/screws should be placed so as to be as divergent as possible from each other and prevent extraction, when tension is possible, and should be straight otherwise (Figure 1.10).

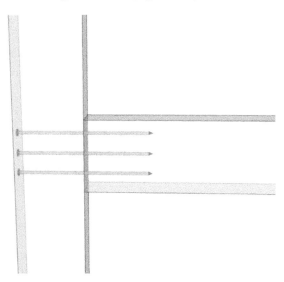

Figure 1.10. *Light-frame end-to-side butt-joint executed with screws or nails orthogonal to the frame components. Splitting of the wood at the head of the joist is a serious risk in this case. For this reason, this connection is not suitable for heavy loads*

The positions of the nails or screws in the connections depend on whether the free end of the member connected at the side is accessible or not. If it is accessible, the nails or screws should be placed from the free side (Figure 1.10); otherwise, they should be placed from the other side, diagonally – and indeed divergently (Figure 1.12). The most typical case is that of the connection of the studs to the wooden plates on top of a foundation wall.

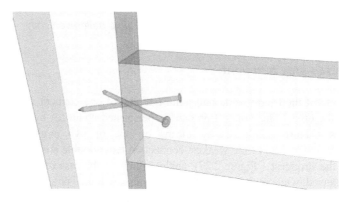

Figure 1.11. *Light-frame end-to-side butt-joint executed with screws or nails positioned diagonally. Because of the position of the screws/nails, this solution is more suitable for resisting vertical shear than lateral horizontal shear*

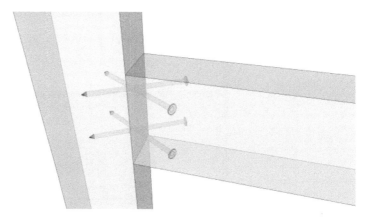

Figure 1.12. *Light-frame end-to-side butt-joint executed with screws or nails inserted diagonally. This solution and the one in the previous figure are less strong than the one shown in Figure 1.10. Because of the position of the screws/nails, this solution, like the previous one, is more suitable for resisting vertical shear than lateral horizontal shear*

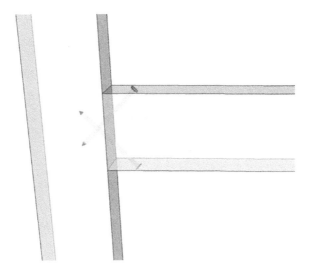

Figure 1.13. *Light-frame end-to-side butt-joint executed with screws or nails placed diagonally. Because of the position of the screws, this solution is more suitable for resisting lateral horizontal shear than vertical shear*

1.2.3. Lap joints

Lap joints are the easiest to implement on-site, because they usually do not require a great cutting precision for the timbers. Indeed, they can be easily made fit to the desired structural dimensions by sliding the elements to be connected until they find themselves in a correct mutual position.

The strength of a simple lap joint depends on how many nails, bolts or screws are used in it, and therefore also on the size (length and width) of the overlaps. The general rule is: the greater the overlap, and the more numerous the connectors (nails, screws or bolts), the greater the obtained shear resistance – provided that the connectors do not interfere with each other. Because the single screws or nails are only resistant to shear, the rigidity of a lap connection (i.e. its resistance to bending or its ability to transmit bending forces) depends on the length of the overlap (provided that the overlapped parts are connected diffusely), and bending resistance is not reached unless the overlap is very long, in the order of 1/3 of the elements to overlap – which is most of the time a situation not worth pursuing in greenhouse structures (Figure 2.32).

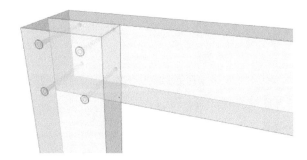

Figure 1.14. *Lap joint between a post/stud/mullion and a beam/joist/rafter. This lap joint features a number of screws or nails sufficient to make it classifiable both as a light-frame connection and as a timber-frame one, depending on the size and strength of both the wooden components and of the screws, nails or bolts*

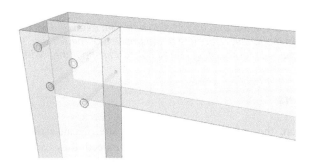

Figure 1.15. *Lap joint between a post/stud/mullion and a beam/joist/rafter. This lap joint is similar to that in the previous figure, but it differs from that, because here, the screws/nails/bolts are placed on different lines along the wood grain, in order to reduce the risk of splitting the wood along the grains concerned. In this example, avoiding the alignment vertically reduces the risk of splitting the wood along the grain in the upright element, and avoiding the alignment horizontally reduces the risk of splitting the wood along the grain of the horizontal element*

1.2.4. *Nailed connections*

Simple nailing is the lowest-cost connection technique and is poorly reversible in practice. Indeed, when a nail is extracted, it cannot be reintroduced into the same lodgment, because the resulting anchorage would be too weak. Nailing can be performed by hitting the nail with a hammer (traditional technique) or shooting it with a pneumatic pistol (modern variant). Pneumatic nail drivers greatly simplify the nailing operations and make nailing very competitive with respect to screwing, due to the lower cost of nails compared to screws.

Similarly to screwed connections, when nailed connections have to resist extraction (tension), the nails should be placed diagonally in directions divergent from each other. The nails can be used directly on the wood, or applied to hot-rolled (stronger) or cold-rolled (weaker) steel connectors, like plates and angles ("L"s), as well as "U"s or "T"s, in a diffuse manner, in order to avoid concentration of stress[2].

Modern variants of the "classical" nails are the cramps and staples (most commonly, U-shaped, Z-shaped or L-shaped) that are suitable for performing different and more complex joining functions than ordinary nails (Figure 1.16).

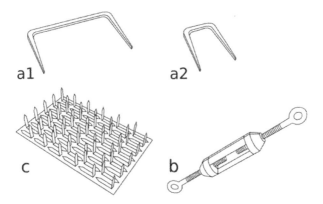

Figure 1.16. *Some types of steel fasteners: a1. cramps; a2. staples; b. tension rod; c. tooth plate*

For nailed connections, steel nails or fasteners or aluminum nails or fasteners can be used. Aluminum nails and fasteners are costlier (as well as, possibly, bulkier) than steel ones, and have the disadvantage of a smaller mechanical resistance, but have the advantage of being suitable for cutting with tools (circular saws, alternated saws) similar to those used for timber in carpentry. For this reason, aluminum-fastened connections have a useful place in wooden carpentry, especially in factories, where they can be used as a lower-impact alternative to diffuse gluing in components to be successively sawn or milled.

On references

The book by Gordon (1978) is known for its effective explanations of the basics of structural behavior.

---

[2] The consequences of the concentration of stress are analyzed with unequalled craft in the book by Gordon (1978), entitled *Structures. Or, Why Things Don't Fall Down*.

1.2.5. Screwed connections

Screwing requires configurations that are identical to those used in nailing, but is reversible: a screw can usually be re-screwed at least one time, and sometimes more than one time, into the lodgment from which it has been unscrewed, and the resistance to extraction of screws is greater than that of nails, especially in the axial direction. These are the reasons why screws compete with nails despite their higher costs.

Screwing is customarily executed with electric screwdrivers powered by batteries and requires fewer skills than nailing with a hammer.

1.2.6. Bolted connections

Bolting in light frames is used only in a small number of cases, when a connection must be particularly strong, or fully reversible, like, for example, in trusses, or when concentrated forces are created by joining several light structural members (e.g. when posts are formed by joining multiple studs together, or when beams or girders are formed by joining multiple joists).

Bolts in wooden carpentry are often accompanied by structural washers, suitable for spreading the stresses onto a broader surface, or use plates, or tooth plates (Figure 1.16). Washers can be wide and thick, in order to be suitable for diffusing the stresses well onto the wood and, together with the bolts and/or the nuts, may be sunk into the element to be connected if it is convenient to keep them flush with the wooden surface – which may happen, for example, to avoid conflicts with the cladding.

1.2.7. Tooth plate connections

Tooth plate (or gusset plate) connections can substitute the connections constituted by plates and screws or nails. Tooth plates are stuck into timber components by using pressure, slow and constant, obtained by means of presses (when the process is automatic) or clamps (when the process is manual). However, when they are cold-formed – which they usually are – they are not very resistant to fire and corrosion. This suggests they should be protected from fire. The tendency to rapid rusting in moist environments, derived from the thinness of tooth plates, also makes tooth plates not very suitable for agricultural greenhouses if they aim to be long-lasting.

1.2.8. *Glued connections*

Glued connections are a broad family, alternative to that of the connections executed by means of steel or aluminum connectors. The most commonly used glues for connecting wooden components in factories (principally for producing glued-laminated timber) are synthetic resins, such as melamine resins, phenol-resorcinol resins and polyurethanes (ureic glues). Melamine resins and phenol-resorcinol resins contain formaldehyde, which is toxic. This is a disadvantage both for indoor usage and as regards the recyclability of the material. The disadvantage of polyurethane resins is instead only related to recyclability, and not to potential health issues derived from the release of substances into the air. Casein, the glue that was used (through a heat-based melting process) for joining timber in pre-modern carpentry, is almost not used anymore in modern carpentry, due to its comparatively low durability, which in turn derives from the fact that, besides being soluble, casein is biodegradable. In the future this may be seen as a feature from the standpoint of sustainability, but that time has yet to come.

The most appropriate glues for connecting timber components on-site are epoxy glues. Epoxy glues have the greatest advantage when the joint to be executed is large, and they can also be used to fill gaps in the wooden components. But because epoxy glues are not UV-resistant, they have to be protected with UV inhibitors.

Epoxies, like polyesters, come in two components: a hardener and a resin, and once set, epoxy is so strong that it can even be used for executing lap joints. And because it bonds well with a variety of materials besides wood – like steel, aluminum and concrete – it can even be used to glue timber components to structural components of those other materials.

1.3. Types of connections between structural sub-systems

Light frames are quite modular, and can be imagined as composed of interrelated sub-systems. The connections involved in the combination of the main sub-systems are the following.

1.3.1. *Interfacing the frames of the front façade and the roof*

The easiest solution for platform-frame greenhouses requires one or two wooden plates on top of the wall frame to complete the frame itself (two plates if the frame is less wide than the whole width of the greenhouse, so as to overlap it on the separated top plates below, and join them). A header joist (fascia beam) on top of it (or them) may be required, depending on the bending stresses, or on the need to facilitate the fixing of the gutters and pressure caps at the eave. As seen, each rafter is usually positioned on top of each stud/mullion (Figure 1.2), especially if the header joist (fascia beam) is weak, absent or not well connected with the other structural members.

A variant of this solution does not use header joists. In this case, two plates may be used in place of one, in order to contribute to the connection between the two framed parts (Figure 1.1). Further variants on this solution require one to connect the plate below the rafters (Figure 2.61) or at the inside face of the studs (or timber-frame posts – Figure 2.63), with the consequence of enabling the connection between the studs and the rafters.

Yet, another variation is keeping the top plate at the very top of the vertical wall structure. This requires the studs/mullions to be full-height. In that case, the rafters have to be connected at the back face of the studs, making that joint resistant to loads, possibly by placing a horizontal plate tilted vertically below it, as a stringer, in a corbel-like (or ledger-like) fashion (Figure 2.65).

The size of the rigid transparent panels dictates the interaxis between the studs and between the rafters. When glass panels are used, the interaxis actually has to be a little larger than that: in the range of 2 or 3 cm less, one per side, in order to (a) avoid that the screws pressing the pressure caps holding the glass panels enter in touch with them locally and create in them an excessive concentration of stresses; (b) leave some free space in the joint between panels, so as to allow for thermal movements and movements due to moisture variations.

In the case of transparent plastic panel enclosures, instead, as seen, the size of the panels can be a multiple of the space between studs (possibly, again, minus about 1–1.5 cm at each side, to leave some space in the joint, when pressure caps are used), because such panels can span continuously across multiple studs or rafters.

In any case, it is beneficial for the resistance of the façade that each transparent panel is restrained at all its sides, and this in turn requires that transoms are provided beneath each horizontal joint. The position of the transoms at the connections between the greenhouse façades and roofs determines the position of the flashing, which has to run again from the horizontal line of the top seal of the building wall to the horizontal pressure cap of the transom below it (Figure 1.28, Volume 3). Laterally, the joints between pieces of flashing have to be overlapped for a length suitable for preventing water infiltration, which is usually considered to be about 10–15 cm. Then, the overlap has to be sealed with silicone, or welded.

The two main configuration solutions in the zone of the greenhouse eave are: (a) positioning the plates/transoms of the façade as high as possible in it; and

(b) positioning the roof transoms as forward as possible in the front. This can be done in several alternative ways: (1) without modifying the transoms (Figure 2.8); (2) modifying the transoms and keeping the roof slope; and (3) modifying the transoms while making them upright. This usually requires sloping their top side with the same pitch as the roof (Figure 2.9).

Similar configurations can be obtained: (a) by placing a header joist (fascia beam) at the edge of the frame, and nailing or screwing the rafters to it with a butt joint (Figure 1.5), or (b) by placing the transoms between the rafters and making the rafters lined externally at the edge (Figure 1.18). The latter solution is stronger in carrying loads, because the studs below the rafters' ends are loaded in a more centered manner, but is weaker in transmitting tension in the direction transversal to the rafters, because the transoms are not continuous, but rather joined with screws that do not even work in shear.

The simplest solution entails placing the two plates/transoms in the ordinary position; but this position requires a larger flashing, at the expense of the potential for solar gains and lighting levels in the greenhouse. A further solution is substituting two plates/transoms with one, worked out so as to support the edges of both the façade panels and the roof panels.

1.3.2. *Interfacing the frames of the side walls and the roof*

The decisions about the connections between the side walls (or façades) and the side edges of the roof depend substantially on the kind of transparent enclosure that is used. When glass panels are used, in the most robust configurations, there is a wooden fascia rafter at the edges of the roof, and the edge rafter is usually sitting on the plate, in line with the external edge of it (Figure 1.18) or the internal one (Figure 1.17, left). Alternative possibilities entail that (a) the rafters have dimensions (and strength) that make it possible to anchor to them both the roof panel(s) and the façade panel(s) (Figure 1.17, right); (b) the side edge of the greenhouse is solved by one rafter for the roof, another one for the wall (Figure 1.17, center); and (c) the plate running along the slope of the roof, and supported by studs of different heights, is positioned at the side of a rafter (Figures 2.8 and 2.9).

In the latter solution, the studs and the plate: (a) may have their external face flush with the external face of the rafters (Figure 1.18); or (b) they may have their internal face flush with the external one of the rafters (Figures 2.8 and 2.9) (in that case, their height is independent of that of the rafters); or (c) they may have an intermediate position. As a last alternative (d), the rafters may be used without a top plate, but rather, one might close the spaces between their ends with wooden stops, so as to allow for different depths of placement with respect to the rafters.

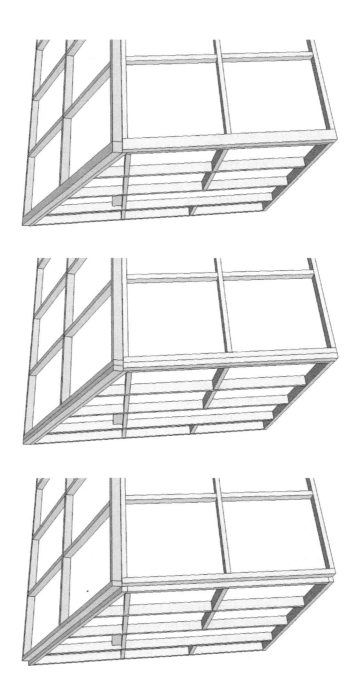

Figure 1.17. *Solutions for arranging the side edge rafters and the side edge plates, alternative to those shown in Figure 1.18*

In all the cases in which the edge rafter is overstressed due to its anchoring role with respect to the side wall (or façade), two (or even three) rafters can be used in place of one, by connecting them to each other by nailing or screwing, so as to make them behave as one. This may be obtained without altering the position of the "master" rafter – the one laid out at the regular interaxis, by placing the "added" rafter externally or internally to the "master" rafter.

1.3.2.1. *Interfacing roof frame and shared wall in attached greenhouses*

In theory, one solution to connect a greenhouse roof and the back wall of a building could be to connect each rafter to the wall individually, with no intermediate elements. But the only situation in which this would not be unpractical is the case in which the building and the roof are built together – both light-framed – and each rafter can be connected directly to a stud. But even in this case, in practice, connecting each rafter to each stud is likely to be tedious, also because performing those connections having the OSB or plywood sheeting panels in between requires precision: a precise correspondence between the position of each stud and the position of each rafter is required. For this reason, the most straightforward solution to connect the rafters to the building studs (or, more generally, to a building wall) requires the mediation of a ledger, constituted by a thick plank, or a joist, laid out horizontally along the building façade at the height of the ends of the rafters (Figure 1.2), or just below the ends of the rafters, as a sort of corbel (Figure 1.3). In the latter situation, to facilitate the connection with the ledger, the rafters are usually worked out at the involved end with a birdsmouth cut (i.e. a cut making the face of the rafter at the support plainly horizontal).

1.3.2.2. *Interfacing the roof frame and the back wall in stand-alone greenhouses*

When the top of the back wall in a stand-alone greenhouse is enclosed with transparent panels, the connection between the greenhouse roof and the back wall resembles very much that of the roof with the front façade. What is likely to be different between the two cases is especially the angle between the two parts (Figure 2.19). And when the bottom part of the back wall is opaque, similarly, the connection between the lower, opaque part of the back wall and the transparent part above may resemble very much that between the transparent part of the front façade and an opaque knee wall (Figure 1.33). When, instead, also the top part of the back wall is opaque, the connection between a transparent roof and the back wall may resemble very much the one that would be between the roof and the front wall in the case in which the front wall were opaque (Figure 1.27); with the difference that, in the former case, the top flashing would protect the cladding of the opaque wall, rather than the pressure cap covering the top transom – very much as it would do if the back wall were transparent (Figure 1.28, Volume 3). The possibility of combination, in that case, regards the position of the roof transoms and the presence or absence of a vertical fascia beam (Figure 1.29, Volume 3) or of a horizontal plate (Figure 1.30, Volume 3).

1.3.2.3. *Vertical joints between greenhouse frame and building*

The vertical joints between the light-frame side (gable) walls/façades of a greenhouse and the façade of a building resemble very much that of a greenhouse roof with a building. An intermediate element – in this case, vertical, a stud – can be used as the last "frontier" of the greenhouse and sealed against the part of the building wall containing the airtight layer (Figure 1.39, Volume 3). That intermediate stud should be anchored both to the top plates and to the bottom plates of the greenhouse frame, and to the structures of the building (studs, posts, solid walls, depending on the situations) via anchors, the type of which depends on the nature of the building wall structure: dowels (in the case of a masonry or a reinforced concrete structure), screws or nails, or bolts (in the case of a wall structure framed with timber or steel).

1.3.2.4. *Sealing the joints*

When an airtight layer is absent from the wall and the wall is massive, the cited intermediate stud should be sealed against the support layer of the wall's plaster (when present), then sealed and plastered at the joints, for the sake of air-tightness (Figure 1.37, Volume 3). In application requiring a high air-tightness, the seal should be constituted by a backing rod plus the sealant, and it may be placed at least at the external edge of the joint – but it could be more cautiously placed at both sides of it: in this way, it produces greater water-tightness and air-tightness and is more durable. But when the execution aims to low cost, or when the intervention is aimed to attach a greenhouse to an existing building, the intermediate element can be sealed directly on the plaster.

The vertical joints may also be protected externally and/or internally from water infiltrations, by keeping the water away from them. This is usually obtained by bending the flashing so as to cover the vertical joints. This solution is rather visible, but sets the basis for greater durability and improved water-tightness.

1.3.2.5. *Relation between the intermediate vertical components and the gable structure*

The mentioned intermediate studs constitute the frontier of the greenhouse with respect to the building and can be constituted directly by the framing elements. In this case, they are going to be more loaded of functions with respect to the other studs/mullions, because they have to perform the same structural action, and they have to be connected with the building structure and walls, and have to ensure additional durability and weather resistance. When this is the situation, those studs should also perform the task of holding and backing the transparent panels, as mullions; which also entails offering anchorage to the pressure caps (if pressure caps are used). In that situation, the joints between the elements may be kept a bit wider to give the pressure caps some extra space for facilitating the assembling operations.

A more complete solution is coupling the last stud/mullion of the gable façade with another stud acting as an intermediary with the façade. In that case, the two studs/mullions should be screwed or nailed against each other, and the inner one may be protected from rainwater by means of flashing (Figure 1.38, Volume 3).

1.3.3. *Gable walls*

Gable walls (side walls) follow the shape dictated by the roof and are therefore likely to have a sloping top edge – except in the rare cases of attached greenhouses that have an almost completely flat roof, or a roof with double slopes and central ridge, or a single-slope roof slanted from gable to gable. Only in those cases, the top edge of the gable walls is horizontal. In the other cases, the top edge of the gable walls is sloped, usually towards the front (but sometimes towards the building).

Each gable wall commonly spans from the lower plate on top of the foundation wall to the side of the greenhouse roof.

As said, usually, the top of a gable wall is formed by a timber plate that follows the slope of the roof and is positioned above the studs/mullions (Figures 1.17 and 1.18). The top of that plate has to be connected to the edge rafter of the roof, which is therefore expected to sustain stresses that are greater than those sustained by the other rafters. For this reason, the edge rafters may be wider than the other rafters (if their section were taller, fitting them dimensionally with the other rafters would be challenging), or may be obtained by joining (screwing or nailing) two rafters (Figure 1.17, center) (or, in some cases, even three).

The vertical structure of the gable walls can be constituted: (a) by continuous studs/mullions of variable height (Figure 1.18) or (b) below, by one order of uniform-level studs/mullions, and above (sitting on top of the part of the wall below, with the intermediation of one or two wooden plates), by a wall portion of variable-height studs/mullions (cripples) giving shape to the upper part of the gable (Figure 1.19).

1.3.3.1. *Light-frame gable wall formed by full-height studs/mullions*

In this solution, each stud/mullion is of a different height, and the transoms between them are placed at regular, homogeneous heights (Figure 1.18). Only the top bays may be exceptions to the ordinary sizes.

The gable wall structure can be designed so as to exert a load-bearing action with respect to the edges of the roof. But in most cases, it is cleaner and safer to provide all the parts of the roof with the mechanical resistance they need to support the roof by themselves, leaving to the gable walls a self-bearing function, and allowing them to move freely with respect to the principal structure under wind loads and mechanical forces of thermal origin.

In the case in which the gable walls are enclosed with double- or triple-layer glass panels, the extra cost of ordering perimeter-sealed glass panels with a bespoke shape can be avoided by cladding the odd bays in question with an opaque envelope, or by closing them with transparent flat multi-layer polycarbonate or acrylic panels, which can be easily cut in a bespoke fashion.

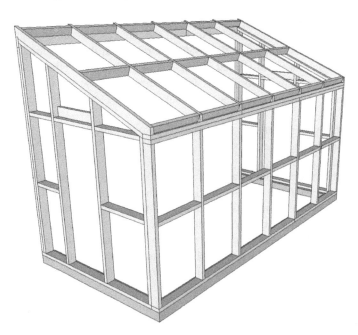

Figure 1.18. *This light frame of an attached greenhouse is complete with frames for the gable walls*

1.3.3.2. *Light-frame gable wall formed by constant-height mullions plus framed top*

It has been seen that the alternative to building the gable wall with full-height studs of different lengths is constructing the lower portion of the wall with studs/mullions of constant length, then surmounting it with a horizontal plate (possibly double, for additional stiffness and strength), that in turn supports cripples of variable height, which finally support a top plate (Figures 1.19 and 2.14). But this solution can be executed as described only when the top part of the gable wall is not tall because, otherwise, a tendency of the wall itself to yield at the structural "hinge" corresponding to the intermediate plate (or plates) would become manifest. That tendency can be contrasted by placing continuous studs in an upright position, spanning from top to bottom, as a backing support of the partial-height studs (Figure 1.20).

When the width of the gable façade is substantial, it is also recommended that the wall framing is divided vertically into two (or more) parts by means of pairs of full-height studs coupled (screwed, nailed) for rigidity. In that case, a solution for making that compound stud stronger is composing it in a "T" shape (Figure 1.26). If any countermeasure for making this kind of gable wall more rigid under load is not taken, this wall type cannot be asked to support the roof load of the roof.

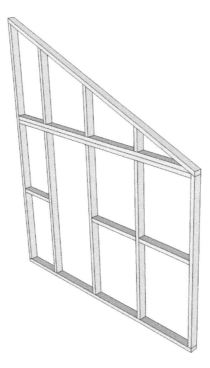

Figure 1.19. *Here, the studs/mullions are discontinuous, and a double plate is laid out onto the ones below. A lintel, in this case, is unnecessary, but the wall is not rigid with respect to vertical loads (because it could form a knee), and is therefore not suitable for load-bearing*

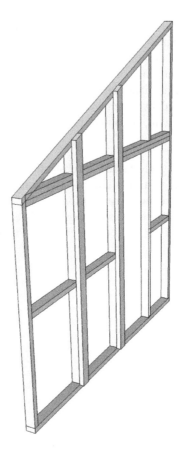

Figure 1.20. *To make a gable wall like that in the previous figure more rigid with respect to the lateral and vertical forces, full-height studs can be added at the back of the partial-height ones*

1.3.3.3. *Possible positions of the gable wall top plate with respect to a roof edge-beam*

The top plate of the gable wall can be positioned in various alternative places. The main possibilities are the following: (a) It may be anchored at the bottom of the outer edge rafter, in such a way that its outer face – and the outer face of the mullions – lies in the same plane as that of the outer edge rafter (Figure 1.22, Volume 3). In this configuration, the gable wall can contribute to supporting the edge rafter, which may allow the use of the latter as a single component (to avoid having to double it). (b) It may be anchored at the side of the outer edge rafter in such a way that the top face of the plate is flush with the top face of the edge rafter

(Figure 1.25, Volume 3). In that case, the upper face of the double plate has to be protected by flashing possibly with rigid insulation in an intermediate position (Figure 1.26, Volume 3). (c) It may be anchored at the side of the edge rafter in such a way that the upper face of the plate is placed at a lower level with respect to the upper face of the edge beam (Figure 1.27, Volume 3). The extreme case is when the double top plate of the gable wall is at the height of the bottom face of the rafter. Also, in that case, the upper face of the double plate has to be protected by flashing. Only that flashing should be shaped to follow the carved-out contour of the side of the roof. (d) It may be anchored to the bottom edge of the outer-edge rafter in such a way that the inner side of the plate is flush with the inner side of the outer edge beam. Also, in this configuration, the top and side of the plate should be protected with flashing. In this case, the studs/mullions may not be placed in a way adequate to support the edge rafter. But when they are, this can make it possible to avoid doubling the edge rafter, so as to have it as a single element. (e) Lastly, there is the possibility that one of the solutions just listed is executed without any wooden plate as an intermediate element between edge rafters and studs. This is possible when the studs are directly anchored onto the edge rafters, rather than onto the top plate. In that situation, the enclosing action may be played by the fascia beam (or header joist) or, less likely, by the transoms between studs/rafters.

1.3.3.4. Doubling the edge rafters of a gable wall

An edge rafter can be doubled in one of the following ways: (a) keeping the *outer* rafter at the regular interaxis between rafters, which entails occluding partially the edge structural bay with the inner rafter. (b) Keeping the *inner* rafter at the regular interaxis between rafters, which entails protecting the outer rafter with flashing, closing on the gasket of the outer pressure cap of the roof. (c) Keeping the interaxis at the joint between the two edge rafters. This requires varying the size of the transparent panels.

1.3.3.5. Cantilevering the roof beyond the gable wall

A further possibility is making the roof cantilever out from the gable walls. But this is a rather complex solution, because it entails connecting some evenly spaced short joists perpendicularly to the external rafter above the gable wall, spanning those joist (outriggers) outwards so as to make it possible to anchor an edge rafter at their ends to the gable wall (Figure 1.23). It is important to note that, to avoid that this solution causes torque in the rafter above the gable wall, the edge rafter should not be supported by the short joists, but anchored to the two header joists cantilevering out from the front and back walls. Or, as an alternative, sloped stops (discontinuous) should be located in place of a continuous rafter on top of the gable walls, and the transoms should be placed at the edge of the frame in outriggers (joists) cantilevering out rigidly from the gable wall. The most common overhang span following this scheme is of a couple of structural bays.

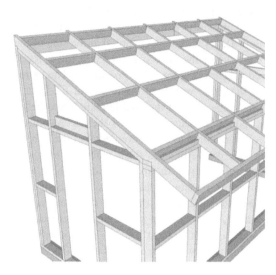

Figure 1.21. In this example, the rafters have been cantilevered out from the front façade and are joined by a header beam

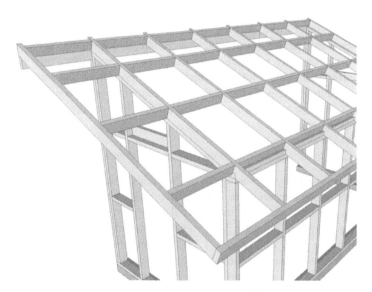

Figure 1.22. In this example, the greenhouse roof of the previous figure has been cantilevered from the gable wall. The edge rafter is supported by the stringer cantilevering out on the back and by the header beam cantilevering out at the front

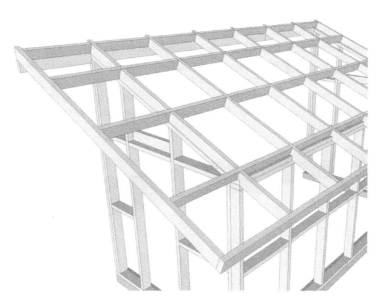

Figure 1.23. *In this example, the edge rafter is supported not only by the stringer cantilevering out on the back and by the header beam cantilevering out at the front, but also by the outriggers at the side of the roof. To make this solution possible, there is no rafter at the middle span on top of the gable wall (except for the segment cantilevering out from the front): there are stops, in its place*

1.3.3.5.1. Joining a greenhouse façade with a ventilated façade or a rainscreen of a building

When the façade of a building to be joined with a greenhouse is ventilated, or when it is clad with a rainscreen, it is necessary to close the greenhouse onto the sheeting of the building wall, where air-tightness resides, rather than onto the cladding, in order to avoid that air circumvents the thermal insulation barriers, and as a consequence, the intermediate layer connecting the greenhouse with the building remains embedded onto the building façade. From this, it becomes clear that artifices must be implemented if the connection is to seem simple, effortless and regular visually, even more than when the façade is not ventilated. The most common artifice is to use an insulated frame as the interface between the greenhouse and the building wall (Figure 1.24).

In that case, the frame in question can be constituted by an upright (stud/mullion) against the wall and another one forming the backing against which the last stud/mullion of the greenhouse is going to be sealed. The space between the two intermediate components has to be insulated and enclosed between panels against which the vapor barrier, internally, can be backed by the panels. Then, the insulation outside has to be brought against the insulation between the two intermediate studs, and the air barrier has to be brought against the gasket of the greenhouse pressure cap to produce air-tightness. The same construction criteria have to be adopted when the end parts of the gable walls of the greenhouse are opaque. In this case, the difference is that the air barrier has to be brought against the external panel enclosing the opaque segment of the wall in question, in order to make it reach the gasket held by the pressure cap. Then, a cladding may be added to that opaque part of the gable wall.

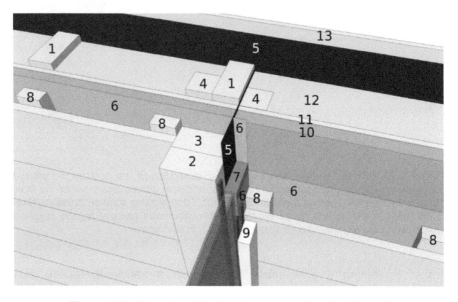

Figure 1.24. *Example of the joint between a glazed gable wall of an attached greenhouse and a light-frame wall of a building*

NOTES ON FIGURE 1.24.– The vapor barrier of the building wall "aims" at the internal joint at the pressure cap, and the air barrier of the building "aims" at the external joint at the pressure cap. An auxiliary, intermediate stud is here used between the mullion of the gable and the concerned stud of the wall, and battens have been added at the side of the wall stud to anchor the battens supporting the wall cladding (wooden planks). The wall is unfinished towards the interior. Legend: 1. wall stud; 2. greenhouse mullion; 3. intermediate stud/mullion; 4. wooden batten for anchoring the battens supporting the cladding; 5. vapor barrier; 6. air barrier; 7. rigid, closed-cell insulation block; 8. wooden batten for anchoring the cladding planks; 9. pressure cap; 10. rigid insulation panel (glasswool, woodwool);

11. sheeting panels (plywood, OSB); 12. insulation bats; and 13. sheeting panels (plywood, OSB).

1.3.4. *Opaque enclosures*

Due to the presence of plants, a greenhouse is a very moist environment. Therefore, the vapor pressure on the inner face of the envelope tends to be very high. This makes the probabilities of condensation within the opaque parts of the envelope high as well. These probabilities can be reduced by laying out vapor barriers at the inside face of the envelope itself, but should also be lowered by making the opaque envelope vapor-permeable in its outer parts. A possibility of escape for the vapor reduces the importance of the integrity of the vapor barrier (which remains, in any case, important).

A way to make the external barrier of the walls vapor-permeable in the most ordinary forms of stud walls, which are usually enclosed by plywood or OSB sheeting, is to create buttonholes (usually of about 5 cm in diameter) towards the outside, in the external wall sheeting. This is necessary because these sheetings are usually not vapor-permeable, due to the glues by means of which they are held together. The buttonholes should ideally be located in the lower parts of the framed cavities, where the air temperatures within the cavities themselves are lower. Placing the buttonholes in a higher position – higher up in the cavities – would create excessive ventilation in the cavities due to the stack effect, which would cause unnecessary thermal losses. In the envelopes enclosed with planks (rather than airtight panels, like OSB and plywood panels) – as well as in any case of envelope characterized by cracks and open joints (in other words, not of an airtight type) – the buttonholes are not necessary.

Air barriers are compatible with buttonholes because they are vapor-permeable, but at the same time, they stop air.

The importance of reducing the likelihood of vapor condensation within the envelope by means of an internal vapor barrier is more important in light-frame walls than in solid, masonry or reinforced concrete walls (a) because of the much greater number of joints they have; (b) because of the greater potential damage that condensation water can cause them, due to the porous materials they are usually composed of; and (c) due to the fact that, despite having the capability to produce a certain moistural "wheel-effect" (Hens 2012c) (i.e. moisture inertia – the equivalent of thermal inertia as regards moisture), this capability is much smaller than that of solid, massive walls.

In constructing opaque envelopes, we should be aware, in the first place, of non-overlapped horizontal joints, because water tends to rest on them, making the components involved decay early. Secondly, we should treat vertical joints with care because, in their case, the force of gravity cannot be used to assure water-tightness. Even when the rainscreen principle is exploited (like, for example, in the astragal configuration – Figure 1.49, Volume 3), it must be taken into account that water runs off by gravity along the joint, and collects at the bottom of it – where it is therefore necessary to expel it by means of suitably shaped flashing or other appropriate projections guiding the water outwards.

On references

The books by Hugo Hens (2012a, 2021b, 2012c) constitute authoritative state-of-the art references about a building-science approach to building envelopes.

1.3.4.1. *Need to balance operable openings to air and thermal losses due to infiltration*

When the openings aimed to reduce the vapor pressure (buttonholes or similar) in the walls are distributed in height, rather than concentrated in the lower zones of the walls, the risk is that stack effect develops in the wall cavities, with the result of increasing air infiltrations and reducing the effective thermal resistance of the overall opaque enclosure. To avoid this, those openings on the external parts of the opaque enclosure should be shielded from winds, in order to prevent the air infiltrated in the insulation-filled wall cavities under the effect of the wind from air-washing the insulation and from causing, consequently, an even more substantial reduction of the insulation effectiveness than that caused by the stack effect. Studies have demonstrated air-washing of the insulation in enclosures built with "dry". Mortarless construction strategy, like ordinary stud walls, is a substantial – and vastly underestimated – cause of reduction of the insulation performances of opaque envelopes well below their nominal performances (Hens 2012b).

1.3.5. *Back walls*

The back wall of a self-standing solar greenhouse can be built in various ways and according to various configurations and thermal functions. To begin with, it can be opaque and massive, opaque and light and insulated, transparent or a combination of those options. Also, it may be opaque and massive below and light and opaque above; or opaque and massive below and transparent above, depending on the design goals. And it can receive the greenhouse frame frontally (Figure 1.25), or it can be enveloped at the side by it (Figure 1.28). Finally, it can be insulated (usually, at the outside face – Figure 1.25) or uninsulated (Figure 1.26).

Most often, the lower portion of a back wall is vertical unless the whole wall is leaned forward, so as to direct solar reflections towards the plants (section 3.7.3.1). And the most usual solutions for the upper part of the wall are that it is vertical as well (Figure 1.34), or sloping away from the equator (towards the north in the Northern Hemisphere) (Figure 1.35). This arrangement allows a tighter spacing between solar greenhouses, when they are laid out in rows. The most thermally advantageous interpretation of that scheme is that of making the upper portion of the back wall, which slopes away from the equator, opaque and insulated, and painted white at the inside face; while the most advantageous solution as regards the amount of natural light admitted is that of making it entirely transparent, so as to admit a greater portion of diffuse light from the sky and cast less shadow backwards.

Several experimental solutions have been attempted, besides the commonest one of a massive back wall as a thermal storage. Among them, there is: (a) the adoption of an actively ventilated cavity wall at the inside face of the back wall (Chen et al. 2018) (Figure 1.37, Volume 1, center), for improving heat transmission; (b) the use of earth-retaining walls as back walls (Zhang et al. 2019) (Figure 1.38, Volume 4); and (c) the use of medium-density components such as straw-based conglomerate blocks for building the back walls (Ren et al. 2019).

As said, the junction between an opaque bottom part of the back wall and a transparent top part, at the construction level, is very similar to that between the knee wall and the part of the front façade surmounting it (Figures 1.33 and 1.34). And the connection between a transparent back wall and a transparent roof is very much like that between a transparent front wall and a transparent roof, except for the fact that, when the roof slope leans towards the front, there is no option for collecting water into channels in the back zone, because the roof slopes away from it (Figures 1.32, 1.33, 1.34, Volume 3).

The joint between a transparent roof and an opaque back wall is usually solved with transoms and flashing, or bringing the wall at the height of the transoms, or bringing the transoms on top of the wall (Figure 1.25).

At all levels of consideration, it is important to take into account that the gable (side) walls and the roof of an attached greenhouse have to be combined with a back wall, in order to allow as much as possible the thermal continuity of the two systems and the air-tightness and water-tightness of the joints.

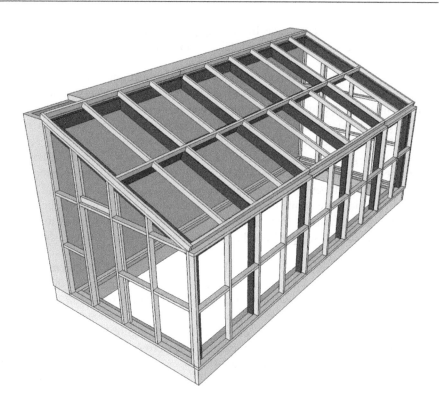

Figure 1.25. *Collaboration of timber-frame (shown in red) and light-frame parts in an attached greenhouse leaning to an externally insulated back wall*

NOTES ON FIGURE 1.25.– In such situations, the insulation of the back wall has to target the frame (ideally, the thermal breaks) in the curtain wall, in such a way that there is no discontinuity of insulation. The reinforced concrete here is shown in gray, and the thermal insulation in yellow. The thermal insulation should "wrap" the wall towards the outside and be continuous with the insulation enveloping the foundation, which is usually constructed gluing water-resistant, closed-cell insulation (like extruded polystyrene, or glass foam) panels onto the foundation, up to the frame in the curtain walls (more precisely: ideally, up to the thermal breaks of the curtain walls).

Light Frames (Wooden Frames) 33

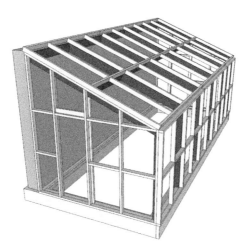

Figure 1.26. *Collaboration of timber-frame (in red) and light-frame parts in an attached greenhouse leaning to an uninsulated back wall. If the back wall is uninsulated, more mass is needed, and the obtained temperatures in winter are lower. In gray, reinforced concrete; in cyan, closed-cell rigid insulation panels*

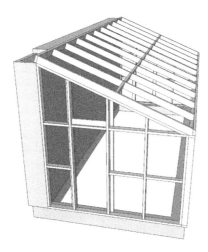

Figure 1.27. *The back wall may be reinforced with steel bars coming from the foundation, or should be thick enough to have resistance to earthquakes on its own, or it should be buttressed, towards the inside and/or the outside. If it is made of reinforced concrete, it should be monolithic with the foundation, in order to cantilever out from the ground safely at the structural level. The case in which the wall has to be buttressed is especially when it is stand-alone and built with masonry. In gray, reinforced concrete; in yellow, thermal insulation*

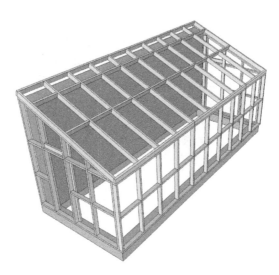

Figure 1.28. To reduce the complication of enveloping the thermal storage wall with insulation, the side faces of the wall may be included in the greenhouse. This allows one to keep the insulation and the cladding of the back wall simpler: flat-shaped, with no corners or reveals. This is a constructive advantage when the wall is closed with an insulated and ventilated façade

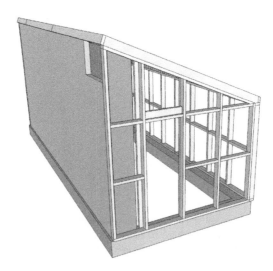

Figure 1.29. The insulation, or the cladding of the insulation, may be positioned so as to be flush with the side (gable) walls. The sides and top of the wall may be left uncovered so as to face the glass or, for higher thermal efficiency, they can be insulated as well

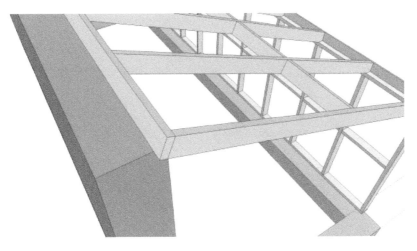

Figure 1.30. *The roof frame can be anchored to the vertical surface of the wall (e.g. via ledgers), or it can be supported by the top of the wall, like in the case shown here*

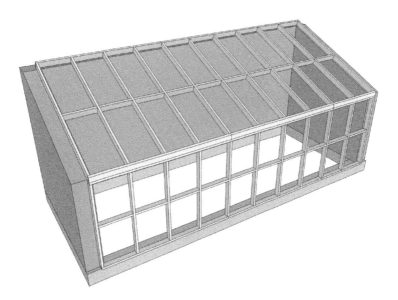

Figure 1.31. *The sides of the greenhouse can be closed by the wall constituting the thermal mass, as shown here, or be glazed, as shown in the previous images. The option shown here is more efficient thermally than the one in the previous figure which, however, keeps the lighting levels of the greenhouse higher*

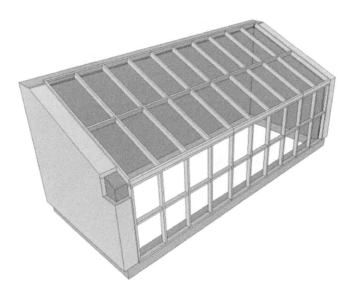

Figure 1.32. *As for the back wall, if the gable walls "wrap" the greenhouse from outside the frames, they should be enveloped in the insulation in such a way to make the insulation continuous with the thermal breaks in the façade and roof. Certainly, this option is an improvement on the previous, uninsulated one*

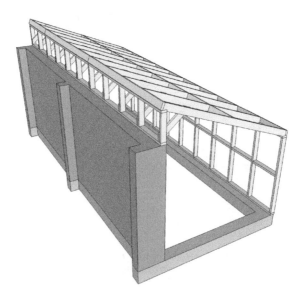

Figure 1.33. *The greenhouse frame can be made higher than the back wall. In this way, cross-ventilation from the back windows is made possible, at the expense of some thermal mass*

NOTES ON FIGURE 1.33.– When the back wall supports the back of the greenhouse frame, it is necessary to brace the greenhouse frame at least along one of the two connections of the rafters (or sloped beams, in the case of timber frames), because otherwise the structural scheme would have four hinges, and would therefore constitute a mechanism. Buttressing the wall to make it safe structurally, as shown here, allows dimensioning its thickness on the basis of the thermal needs rather than of the structural needs.

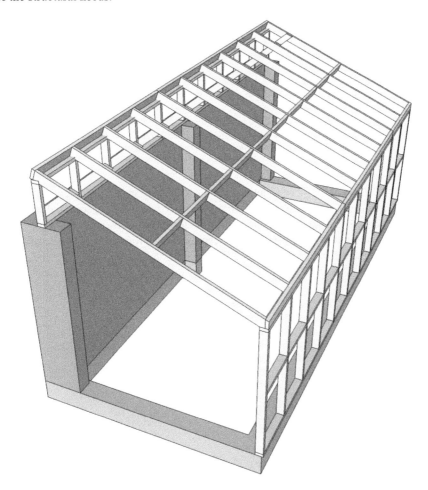

Figure 1.34. *Buttressing the storage wall towards the inside rather than towards the outside is likely to simplify the wall's insulation at the outside face. This structural scheme, as well as the one shown in the previous image, is also suitable for thicker and taller greenhouses*

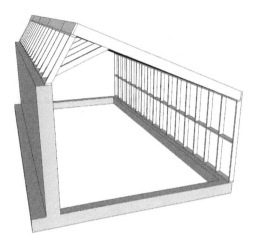

Figure 1.35. *In this scheme, the framed, top part of the back wall has been sloped forward*

NOTES ON FIGURE 1.35. – This solution is in the technical "style" of Chinese agricultural solar greenhouses (where the top, sloped part of the back wall is opaque and insulated) and produces the advantage of casting a shorter shadow towards the back of the greenhouse and reducing the spacing between successive rows of greenhouses, so as to exploit more intensely the available land (Figure 1.50, Volume 1).

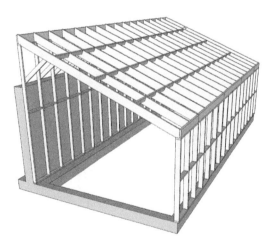

Figure 1.36. *Constructing a back frame independently from the massive wall simplifies the structural requirements of the wall itself and its construction, without renouncing to profit from the wall mass (provided, of course, that the frame against the wall is not glazed)*

On references

Studies exploring the consequences of specific design choices regarding the back wall in solar greenhouses are Zhang et al. (2019), Haoshu Ling et al. (2018 – focusing on a ventilated wall integrating phase-change materials) and Ren et al. (2019 – investigating the consequences of a straw-bale north wall).

1.3.5.1. *Partially transparent back wall in an attached greenhouse*

When the top back eave of an attached greenhouse is higher than the top front eave of the building, the shared wall must be completed by a transparent or opaque back top, which may or may not host operable openings in it (Figure 1.37). In these cases, the structure of the top back façade of the greenhouse can surmount the structure of the building, or be independent of it (Figure 1.36). This situation can be used to improve the passive ventilation potential in the greenhouse.

When the structure of the building supports the structure of the top back façade of the greenhouse, the top back façade can sit on the top front edge beam of the building as it would sit on a foundation wall, aiming to give continuity to the slopes of the waterproof layer of the building roof, so as to avoid the occurrence of any joint in a position lower than that at which rainwater is collected. But when the parapet of the building is not strong enough to support the greenhouse back, the structural connection has to take place beyond it. As an alternative, the parapet may be avoided altogether to allow for a stronger connection.

When it is possible to cantilever a support-beam/fascia-beam from a beam of the building, the studs/mullions of the top back façade of the greenhouse can also be supported individually by that beam – by sitting on it, or by being anchored in front of it.

As an alternative, the structure of the top back façade of the greenhouse can be built as an independent frame sitting on its own foundation support (Figure 1.36), connected to the building structure, a bit as if the latter were a bracing device. Of course, in that case, the foundation front wall of the building should be up to the task of supporting the greenhouse; otherwise, additional foundations should be created to support the greenhouse studs or post of the back wall.

In the described situation, both in the case in which operable openings at the back end of the greenhouse are absent and when they are present, the top back façade of the greenhouse can be made opaque and insulated or transparent. In the case in which it is opaque, its openings in turn can be constructed as opaque or transparent. The opaque solution is more advantageous at the solar thermal level and at the level of thermal stability, but the transparent solution produces higher lighting levels, at the cost of lower temperature averages in winter, as well as of wider thermal swings year-round. The better solar thermal performance of the opaque solution derives from the fact that the transparent solution does not prevent solar radiation from exiting the greenhouse unabsorbed, after entering it.

Figure 1.37. *Design example of an attached greenhouse that is taller than the building. Design by the author*

NOTES ON FIGURE 1.37.– The openings at the rear top of the greenhouse encourage stack-effect ventilation, and, for some wind directions, wind-driven ventilation as well. The curved trusses in this scheme (experimental) are constructed with lap joints. A central massive compressed earth wall is shown in blue. Both the wall and the soil in the greenhouse exchange heat with the ground below the greenhouse, but are insulated away from the surrounding superficial ground. In the image at middle-height of the right-hand side: view of the foundation arrangement: it can be seen that the greenhouse floor and the basis of the central massive wall are insulated at the perimeter, but not against the ground, and the northern half of the building, adjoined to the shared wall, is a superinsulated wooden frame on low stilts made of concrete hollow blocks. This arrangement is conceived to make possible the seasonal exploitation of the thermal bulb into the ground, as well as the daily exploitation of the mass of the shared wall (without entailing thermal bridges), despite the lightness of the construction method used for the house and its being detached from the ground. Below, in green: the thermal bulb obtained in the ground (i.e. idealized shape of the earth mass involved in the seasonal thermal storage). Given the lack of openability of the arched front of the greenhouse and the reasonably short width of the greenhouse, the radical stance taken here is to pursue ventilation on the basis of a complete opening at the sides and at the top back of the greenhouse. The horizontal timbers on the curved façade work as anchors for the seasonally movable shading devices, which are meant to be constituted by tensioned canvases.

There is also the possibility of extending an attached greenhouse sideways beyond the façade of the building. This may take place at one side of the greenhouse or at both sides. The reason for this solution may be to produce more solar gains for the building (this makes sense especially in the case in which the added solar space is connected thermally to the building via a convective heat transfer strategy), and/or creating more room in the greenhouse – for example, for growing plants. The free side portion of the back wall of the greenhouse in those cases may be opaque and insulated or may be transparent, and the openings integrated into it may also be opaque or transparent. In the case in which the side portion of the back wall is opaque, it should be checked that it does not obstruct too much solar radiation from the side walls of the building.

As in the case of the other back walls, the advantage of an opaque side-back wall is that it is more efficient thermally with respect to heating, and the advantage of a transparent one is that it is suitable for producing higher daylight levels.

At a construction level, the joint between the back wall of the greenhouse and the side walls of the building can be similar to the described one between the side walls of the greenhouse and the front wall of the building (see Figure 1.24, and Volume 3, Figures 1.32, 1.33, 1.34).

1.3.6. *Connection between the roof of a building and the roof of the greenhouse attached to it*

The transparent panels of the greenhouse roof can be positioned at the same height as that of the building roof – in which case, the waterproof layer (usually constituted by flashing, which in turn is overlapped with the waterproof layer) overlaps with the transparent panels (Figures 1.35 and 1.36, Volume 3) – or they can be placed below the level of the building eave, in order to make the flashing overhang on the highest transom and protect the joints from rainwater.

1.4. Alternative structural solution: light-frame curtain walls supported by horizontal purlins

The solutions based on mullions and transoms in which the span of the transoms is longer than the span of the mullions (as well as its variant in which purlins are used in place of the transoms and stops are placed in front of the mullions, in order to make them flush with the purlins) are seldom used with wooden frames because, due to the span of the transoms, the transoms would have to sustain substantial forces, and would therefore require a substantial resistance to bending and rigidity against sagging. A solution for addressing this issue without making the wooden frame too bulky is that of giving the transoms a deep (stud-like or rafter-like) section, adequate to stand greater bending forces. However, this solution is also likely to reduce the solar access, and therefore the solar gains, especially if the transoms/purlins are not detached from the glass panels by spacers.

Another problem related to combining long transoms or long purlins with mullion/studs/rafters is that when the mullion/studs/rafters are of standard sizes, they, being few, may be heavily loaded. Certainly, this may be addressed by making them stronger, for example by using, in their place, compound posts/beams formed by two or three studs/rafters.

Due to the substantial bending of long-span transoms/purlins under load, the constructional schemes entailing long transoms, or long purlins, are more often used in combination with plastic panel enclosures, because multi-wall plastic panels are much lighter than multiple glass panels, and endure deflection better. In that case, the plastic multi-wall panels are often made continuous along the horizontal structure. The described solution is frequently used in combination with portal frames, trussed or non-trussed, in which case, purlins spanning from portal to portal are often used in place of transoms spanning from stud/mullion to stud/mullion (Figure 2.78).

1.5. Alternative structural solution: trussed light-frame structures

In this solution, both the studs and the rafters are substituted by trussed frames and portal frames (or semi-portal frames). The solidarity between the uprights and the beams of the portal frames can be obtained by making the structure continuous – which happens when each upright of the portals "flows" into its beam in a continuous scheme (Figures 2.76, 2.77 and 2.78) – or, alternatively, when each upright and its beam are connected rigidly – for example, with a lap joint (Figure 2.79), or with a butt joint suitable for resisting the bending force. However, the use of portal frames of the light-frame type spaced at short interaxes is costly, and therefore most of the time, the spacing between trussed light portal frames is greater than in non-trussed light portal frames, and more similar to that of timber portal frames, even when the trussed frames are built with timbers of small section, typical of light frames (Figures 2.76 and 2.79). This is the reason why, in this work, the structures derived from trussing light-frame components, rather than being addressed in this chapter, dedicated to light frames (tightly spaced), will be addressed in the next chapter, dedicated to the more widely spaced timber frames (see section 2.3.3.2). In fact, trussed frames demonstrate more than any other type of frame that the line distinguishing light frames from timber frames is blurred.

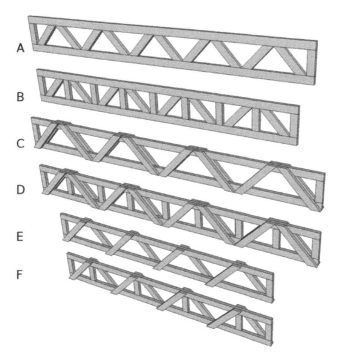

Figure 1.38. *Types of wooden truss beams*

NOTES ON FIGURE 1.38. – A. With diagonals and no uprights, butt-joined. Profiting from the fact that the diagonals and the chords are flush, often this configuration is executed with tooth plates at the sides of the woods. B. With diagonal and uprights, butt-joined. This solution is stronger than solution A, because it is more hyperstatic. C. With diagonals and no uprights, lap-joined. This solution is easier to build than solution A, but it requires more wood unless plates are used for the diagonals. D. With lap-joined diagonals and butt-joined uprights. This solution is more hyperstatic and stronger than solution C. E. With lap-joined diagonals, asymmetric, with no uprights. This is the quickest solution to build, as well as the least strong, given the increased risk of side buckling. F. With lap-joined diagonals, asymmetric and uprights. Stronger than solution E.

On references

The design of trussed timber beams assembled with lapped joints had a golden period in the 1950s and 1960s, but its position on the market was gradually challenged by the appearance of large-span glulam beams. A very good book on the topic of timber-framed trussed techniques is by Timber Engineering Company (1956).

1.5.1. *Trusses and trussed portals or semi-portals orthogonal to the front façade*

The possible geometric combinations of modern wooden trusses are many, and the most common layouts for trussed portal frames is by far those in which the portal frames are orthogonal to the greenhouse front. The fact that, in portal frames, the connections between each beam and its uprights are rigid allows a lot of freedom in their overall geometry. The two principal arrangements of the trusses in wooden truss portal frames are (a) that in which the elements – chords, diagonals and short uprights – are connected with lap joints (Figure 1.38(C–F)); (b) that in which the elements are connected along the same plane with butt joints (Figure 1.38(A) and (B)).

1.5.2. *Light-frame trussed portals parallel to the front façade*

The solution of light-frame trussed portal frames or semi-portal frames parallel to the front façade (Figure 2.84) is seldom used, mainly because of the greater shading effect of the portal frames when they are oriented perpendicularly to solar radiation rather than along its direction. An advantage of making the portal frames orthogonal to the front façade is that the elements of the secondary structures (e.g. purlins) in that case are parallel to the slope direction, which is advantageous for fixing the transparent panels. Indeed, in this way, the panels can be positioned with the longer side along the slope direction – which is safer, because the so-formed longer joints can be more easily protected by pressure caps.

1.6. Criteria for the construction of light-frame trusses and trussed portal frames

1.6.1. *Chords, diagonals and uprights overlapped on different planes*

When chords, diagonals and uprights overlap on different planes, in order to keep the configuration symmetric along the principal axis, the connections can be executed by doubling the diagonals of the trusses, and possibly the uprights (Figure 2.76), or by doubling the upper and lower chords. In the former case, if all the elements are constructed using the same wood section sizes, it is likely that the diagonal will come out oversized. A solution for addressing the oversizing issue is to use wider and flatter wooden planks in place of battens for the diagonals (Figure 3.4, Volume 4). When the planks forming the diagonal are in compression, they have a lower buckling resistance than the battens (amount of material being equal), but, on the other hand, the connections at their ends are more rigid, thanks to their greater lever arm. This may be overall an advantageous arrangement. Another solution is oversizing the chords, or coupling them two by two (Figure 2.76), or three by three – considering that the stronger the portal frames are, the larger the interaxis between them could be. In the case in which the span between the portal frames is substantial and glazing panels are going to be used as transparent enclosures, the glazing may be placed with the long edge transversal to the slope; but in that case, there is the disadvantage that many panel-to-panel transversal joints (which are more difficult to make watertight than the joints along the pitch) would be necessary.

From the point of view of how easily the transparent panels can be installed, the solution in which the chords are single and the diagonals and the uprights are double is easier to work out, because the transparent panels will be fixed onto the portal centers (Figure 3.4, Volume 4).

1.6.2. *Chords, diagonals and uprights on the same plane*

When the chords, the diagonals and the uprights lie on the same plane, the connections are all constituted by butt joints, and their execution is likely to require high precision. And also, when the trussed columns of the portal frames are built in the same plane as the trussed rafter, the inner of the two chords of the trussed rafter, or of the two chords of the upright, must be interrupted. This can be addressed by doubling the chords or the uprights and diagonals of the trussed beams, and assembling them by lap-joining them, so as to make the beams symmetric along the longitudinal axis (Figures 2.76 and 3.8, Volume 4), and suitable for incorporating the columns in the place of some uprights. An alternative is to keep the columns and the rafters on two different planes, and overlap them so as to make the post-and-beam arrangement asymmetric (Figure 2.79).

Regarding the connection technique to be used: when the components of the elements are on the same plane (i.e. when butt-joints are used), nailing or screwing the timbers along the grain should be avoided. For this reason, these connections are most commonly executed with the aid of cover-plates (helper side-plates, so to speak). One solution is to use tooth plates pressed in place; but the thinness of tooth plates exposes them to the risk of having a too short life span in the humid environment of the greenhouse, due to corrosion. Another solution is using wooden cover plates – or wooden battens – in place of light-gauge steel plates; but this – especially when battens are used – makes lapping the frames at the connections unpractical.

Of course, both trussed rafters and trussed columns can have slopes, and their chords may not be parallel.

1.6.3. *Transversal connection of portal frames*

The most straightforward solution for connecting the portal frames transversally is based on horizontal purlins. In this scheme, the purlins work like distributed secondary structural elements – a bit like beams distributed throughout the portal frames (Figure 2.77). A solution for limiting the size of the purlins without compromising their bending resistance is coupling them two by two with a space between, and connecting them with each other in a shear-resistant fashion with the mediation of intermediate blocks (so as to increase the overall bending resistance and torque resistance of the arrangement) (see example given in Figure 3.4, Volume 4).

1.6.4. *Bracing strategies in light frames*

Bracing against lateral forces is only unnecessary in very small greenhouses, where the bracing function can be performed instead by the rigidity of the connections. In all other cases, bracing is necessary on the vertical planes along the two orthogonal principal directions of a greenhouse, and on the plane – or planes – of the roof.

In a light-frame greenhouse, bracing is usually pursued by means of full-length (whole-structural-bay-length) diagonals (Figure 2.111), possibly spanning more than one structural bay. This solution is used especially when the bracing elements are to be integrated into the transparent parts of the envelope, in empty structural bays (modules) – that is, in plain sight. The bracing components used can be wooden diagonals or diagonals constituted by tensioned steel cables or steel rods (Figure 2.105).

The principal alternative is to use systems of diagonals spanning one structural bay (module) each, especially when the bracing elements can be hidden in an opaque part of the envelope, or in an opaque cladding within a structural bay (Figures 2.106 and 2.108). When opaque enclosures are involved, a heavy but conceptually simple alternative is bracing the structure with rigid panels (e.g. framed panels made rigid by OSB or plywood sheetings, or X-lam panels) in place of the diagonals.

In the case of light frames, the solution of using partial-length butt-joined diagonals as bracing, which is typical of traditional timber frames (Figures 2.26 and 2.104 (last three columns)) is most often avoided, and rather, the lap-joined solution is used.

In small- and medium-sized greenhouses, bracing is usually required on the greenhouse enclosures (façades and roof) (Figure 3.14, Volume 4), while in large greenhouses, some additional bracing is usually also necessary along the internal structural bays.

The extension and configuration of the bracing arrangements depend on the configuration of the trussed structure in question, its dimensions and the lateral loads due to wind which the structure may have to resist. The minimum result that the bracing diagonals have to obtain is to reduce the degrees of freedom of the structure so that it cannot be displaced; but usually, they go much beyond that. Indeed, a certain redundancy – generating hyperstaticity – is beneficial to the built solutions, because it makes them safer and more robust.

In attached greenhouses, the greenhouse structure, besides being braced in itself, should be anchored to the structure of the building, so as to avoid that, under the thrust of horizontal forces, the two can move in a way inconsistent with each other, up to the point of colliding with each other.

1.6.4.1. Wooden diagonals as bracing elements

It has been seen that the position of the bracing diagonals can be arranged (a) so as to create a truss configuration across the full extension of the greenhouse (Figure 2.111) – which makes the collaboration between different bays indispensable for the propagation of the bracing resistance across the framed structure; or, as an alternative, (b) some structural bays, distributed so as to make their influence involve the whole structure (possibly, positioned symmetrically, and with the criterion of "closing" their layout), can be fully braced in themselves, with respect to both the possible verses of the horizontal force along its plane (Figure 2.106).

In the framework of the former solution (solution a), it is better to make the scheme of the diagonals continuous in a zig-zag fashion rather than arranged in a more localized fashion, because this makes it possible to distribute the stresses and the deflections more evenly (Figure 2.106). The solution of diffusing the bracing arrangement to the whole structure is also suitable for allowing the use of partial-length diagonals, thanks to the spreading of forces that this solution entails (Figure 2.111). The rationale of the latter solution (solution b) is instead making the structure more rigidly braced in fewer zones.

When the localized solution substitutes a distributed one, X-configured bracing schemes are likely to substitute V-configured ones. And when a distributed frame conception is X-braced in the first place, the concentrated bracings are likely to have to be constituted by stronger X-bracings.

The connection between a diagonal member and each structural bay light-framed internally (to make the transparent enclosures installable), besides being executed to the primary structural elements of the bay, usually is also distributed to all the secondary structural elements within the bays (mullion/studs, or, joists, or rafters, or purlins). The solution of cutting through the studs/mullions/joists/rafters/purlins with the bracing diagonals, which is common in the case of traditional platform framing, is indeed not common in the case of wooden, light-framed greenhouses, due to the fact that the weakening of the mullions/studs entailed by the operation seems usually excessive in a situation like that of transparent bays of greenhouses, where sheeting panels are absent.

In the case that the studs/mullions/joists/rafters/purlins of a light frame are very widely spaced – which may be, for example, when a greenhouse structure is built with light portal frames (Figure 2.75), or with portal frame trusses (Figure 2.78) – the connections of the bracing diagonals with the studs/ mullions/joists/rafters/purlins can be lapped (by screwing, nailing or bolting, with or without ring connectors[3], depending on the expected forces in play) or can be butted; but lapping them makes the connectors (screws, nails, bolts) work in shear without requiring plate connectors or steel angles, and is therefore simpler. In the case in which the connections are lapped, they can be expected to work both in tension and in compression, which explains why "chained", continuous configurations of the diagonals of the type just described (the distributed connections at the secondary elements) are often preferred: by reducing the free span of the diagonals, their

---

3 In light frames, usually without ring connectors.

resistance to bucking in compression is much increased, and the hyperstaticity (as well as the robustness of the structure) is increased as well.

Another consequence of the lap joints is that, because lap-joined wooden diagonals work well both in tension and in compression, a choice is possible between arranging the diagonals in an open and distributed fashion, and in a symmetrical and distributed fashion.

In the case in which the connections are, instead, butted, the diagonals can be expected to work only in compression, as struts. This fact may advise the adoption of a symmetrical bracing arrangement based on St. Andrew's crosses ("X" schemes), which, however, requires that the diagonals are given a thickness of half the studs at a maximum, or that they are arranged in a half-wood configuration (Figure 2.48) – and these are inevitably less structurally efficient than full-wood arrangements with respect to the quantity of wood they use.

Executing a butt-joint for a partial (small-sized, partial-bay-length) diagonal requires shaping the head of the diagonal elements so as to make them fit well with the other structural members, and can be obtained by simple axial screwing or nailing; but, due to the fact that the strength of this connection is not great, it is more often used for the framing of windows (Volume 3, Figures 2.26, 2.27 and 2.28), doors and skylights than for whole greenhouse frames.

1.6.4.2. *Cables or rods as bracing elements*

When the bracing diagonals are constituted by cables or rods, they can only work in tension, and therefore their layout in the structure should be defined considering (a) that the cables should be arranged in a symmetric manner, which will likely also be rather strongly hyperstatic; and (b) that in the context of hyperstatic bracing, especially in light frames, typically, "X" configurations (St. Andrew's crosses) are safer than "V" ones. The latter configurations do work well, but are likely to create the conditions for a less homogeneous distribution of stresses across the structure, because when one of the two diagonals of a "V" scheme is in tension, the other one is in compression (Figure 2.104). This makes the structural components anchored to the vertex of the "V" submitted to bending; while in the case of the "X" configurations, a diagonal in each bay is always at work (indeed, in tension) (Figure 2.104). This beneficial redundancy is even greater when the diagonal timbers are also connected with each other at the center of the "Xs".

The position of a cable or rod with respect to the framing members can be lapped or included in the plane of the structural bay. When the position is lapped, the connection at the ends of each cable or rod can be executed by means of a tension rod (this is the simplest, but weakest, solution due to the smaller number of screws or bolts that it allows – one at each cable's end, preferentially working in shear) or by means of a plate, in its turn usually screwed or nailed to the framed elements (again so as to make the screws or bolts work in shear). When, instead, the bracing diagonals are included in the plane of a structural bay, only solutions featuring connectors shaped to transform the tension in shear at the level of the screws working within the timbers are possible. But in the case of bolts, making them work in tension, combined with large and strong washers, is acceptable.

The substantial advantages of the bracing arrangements using cables or rods are (a) simplicity of execution; (b) light-weight; (c) the fact that, thanks to their thinness, they obstruct less solar radiation than wooden diagonals; and (d) the possibility of keeping fine-tuning the tension in time, so as to respond not only to the immediate structural necessities of the structure when built, but also to the necessities arising with time, as the wooden structures adjust themselves reacting to the environment. In wooden greenhouses, due to the adjustment of the frames in time, it may indeed be necessary to re-tune the tension of the bracing diagonals periodically (with a frequency in the order of a year, or two times a year) by acting on the tension rods (Figure 1.16) – which are usually placed at one of the ends of the cables/rods (Figure 3.9, Volume 4), or sometimes at mid-span. The tension rods are usually connected to the cables by means of hooks and rings, and the tension can be adjusted by rotating them along their thread.

1.6.4.3. *Bracing with concentrated rigid panels*

A further alternative for bracing a light-frame greenhouse is making some parts of it rigid by means of rigid panels suitably localized in some structural bay (module). Such panels may be framed panels braced by diagonals or stiffened by exterior plywood or OSB panels, or, when particularly high resistance is needed, may be massive cross-lam panels.

Because panel solutions work decently well for resisting bending moments relative to both rotation verses onto a plane, and both in tension and compression (especially when thick, 2.4 cm plywood or OSB panels are used), the placement scheme of the panels does not necessarily have to be symmetrical. But symmetrical schemes will be in any case more reliable in a wider range of circumstances, and are therefore preferable.

The substantial drawback of the panel solution in greenhouses is that when the panels are parallel to the façades, they obstruct solar radiation. A solution for limiting the reduction of the greenhouse effect caused by this is constituted by detaching the panels substantially from the glazing, so that they cannot entrap the warmed air between themselves and the glazing.

1.7. Intermixing parts of timber frames into light frames

Intermixing in some parts of a greenhouse wooden frame conception and a timber-frame one is possible and useful, especially when particularly demanding (or at least specific) structural tasks are required, of a kind likely to produce heavy localized stresses. An example is that of a frame having different structural spans in different parts. In those cases, where the structure will be called to sustain additional stresses, it could be made stronger (indeed, by constructing it as a timber frame), and connected with the light-frame parts.

Like in a platform-frame structure, some structural elements (e.g. girders or beams forming a central spine along a building, or systems of columns and beams or lintels, making it possible to locate large openings in a façade) can be strengthened up to the point of acquiring a timber-frame-like nature. In this respect, the situation regarding the façade of a building can be quite similar to that regarding the façade of a greenhouse. A similar situation is constituted by intermediate longitudinal beams or girders within a roof structure, dividing the span of the rafters or purlins.

In all cases, whenever a timber-framed part of a structure (constituted of one or more structural bays, for example) is integrated into a light frame, the bracing required by the timber-frame part is of the strong kind suitable for a timber frame, not of the light and distributed kind suitable for a light frame. And the structural elements to which the bracing elements are anchored should be sufficiently strong for the task as well. The objective of "including" a timber-frame sub-system in a light frame indeed often requires one to anchor the bracing elements, ultimately, to the foundation wall, which is often the strongest anchoring resource within the light-frame system.

The simplest case of integration is when the members of the timber frame and of the platform frame are alternated (Figure 1.39). But under some viewpoints, this case only appears to be simple. Indeed, when the light-frame elements and the timber-frame ones are of different widths (the studs/mullions with respect to the posts and the rafters with respect to the sloped beams), and when there is a constant interaxis from center to center between them, the span between them cannot be constant; and vice versa, when the span is kept constant, the interaxis from center to center cannot be constant. This requires variations in the width of the glazing panels or in the width of the pressure caps (see Figure 1.39).

The described modular conflict requires a choice between keeping the interaxis constant, accepting that the span between and studs/mullions (or between beams and rafters) is smaller than that between stud/mullion and stud/mullion (or between rafter and rafter). But sometimes, this modular conflict is confronted by making the span between post and stud/mullion (or between beams and rafters) the same as that between stud/mullion and stud/mullion (or between rafter and rafter), accepting that the interaxis is not constant.

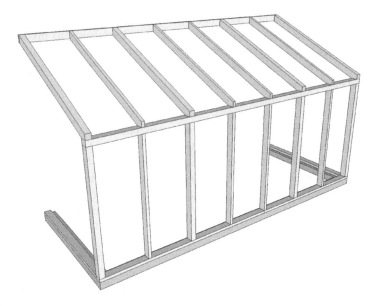

Figure 1.39. *Simple case of intermixing of lighter elements (light frame) and sturdier ones (timber frame), in the case of constant interaxis and therefore non-constant width of the glazing bays. In this example, one every two studs, and one every two rafters, is sturdier than the others*

Intermixing light-frame and timber-frame elements leads to a more difficult execution than keeping the two orders distinct, but requires less construction materials and fewer construction components. Both choices have their advantages and disadvantages.

1.8. Analogies with cold-rolled light frames

It is interesting to note that most of the techniques that are valid for wooden light frames are also applicable to cold-rolled steel light frames (light-gauge steel frames). In most cases, each wooden light-frame solution has a correspondent light-gauge steel, steel light-frame solution, meaning that it can be "translated" point to point, literally, "word by word", into it. This also entails that it is possible to intermix wood and steel light frames, and also that it is possible to substitute one for the other rather interchangeably. However, especially in the case of greenhouses, the much greater thermal resistance of wooden light frames over light-gauge steel frames gives the wooden solutions a substantial advantage. To this disadvantage of light-gauge steel frames, the fact that their thinness put them at serious risk of defaulting by corrosion should be added.

1.9. Arched and vaulted construction in light frames

Arched construction is especially used in combination with curved load-bearing structures, like bowed timbers and bowed metal pipes, and is therefore more common in light configurations; for example, in hoop greenhouses built with steel, bamboo or thin wooden arches (see also section 4.7); but exceptions exist, like the solutions based on glued laminated arches, or on curtain-wall-like configurations arranged in rigid polygons (as shown in Figure 1.40).

Figure 1.40. *Vaulted greenhouse adopting a triangular curtain-wall-like configuration in the Crossrail Place Roof Garden at Canary Warf, London. Photo: sludgegulper, 2016, License Creative Commons*

1.9.1. *Lamella vaults*

So-called lamella vaults were invented by an inventor called Friedrich Zollinger in Germany in the first years of the 1900s, then patented by different persons in several countries. This patenting slowed down their usage; but the patents expired many years ago, and now there are no more obstacles to their widespread application.

The conceptual beauty of lamella vaults is that they can cover large spans by means of short components. This is possible because the components (timbers, in our case) are arranged in such a way that each newly added component during construction cantilevers out about halfway from the components put in place before it (Figure 1.41). Indeed, the lamella vault is a type of reciprocal structure, that is, a structure whose components aid each other reciprocally. In the lamella vault arrangement, each timber works in bending and compression (possibly, tension, due to the wind uplift), but the connections between timbers do not have to be bending resistant. What is obtained is a lattice framework composed of parallel diagonal arches laid out in two mutually intersecting diagonal directions, so as to form bays of rhomboidal shape, and suitable for forming barrel vaults.

Figure 1.41. Lamella roof in a barn interior at Gut Garkau Farm, Germany, 1923–1926. Designer: Hugo Häring. Photo: seier+seier, 2008. License Creative Commons

The lamella vault scheme may be seen as the framed equivalent to the herringbone brick pattern in masonry vaults and domes, of the kind that allows one to build unreinforced vaults and domes without scaffolding. The construction of lamella roofs does need temporary scaffolding, however, which, on the plus side, can be partial, can regard a small part of the total small surface, and can be moved as the construction advances. The scaffolding has to have the same arched shape as the vault to be constructed and has to be placed at its intrados before construction. In this scheme, the roof can be built in "segments": below each "segment", in turn, the scaffolding is placed until completion, then lowered a bit, slid forward of one segment, and hitched up again to support the next construction phase, and so on until completion.

In this scheme, if the shape of the components is straight at the estrados (longitudinally), the vault shape that is obtained is not actually a curve, but a polyline. To obtain a true curved vault from straight timbers, timber shaped to curve profiles at the estrados are needed (Figure 1.42).

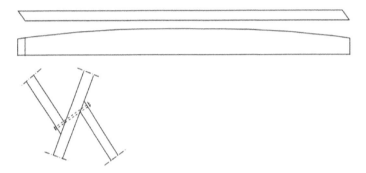

Figure 1.42. *Above: profile of a lamella seen from above and from one side (note the curved extrados). Below: a node between lamellas*

Figure 1.43. *Large-sized lamella vault in Berlin, by Elite Holzbau GmbH. The vault is braced by longitudinal cables. Photo by Kulturagent, 2020. Creative Commons License*

Lamella vaults are frequently covered by opaque structural panels (or planks) bracing the vaults themselves along the tangent plane, but this solution can only be used partially in greenhouses because they need prevalent transparency rather than prevalent opacity. Bracing a lamella vault when most of the enclosure have to be transparent can be obtained in at least two ways. One is making some of the rhomboid modules rigid by closing them with rhomboid panels (usually screwed or nailed to the lamellas, in order to make the parts involved collaborate rigidly), distributed with a frequency and geometry suitable for making that stiffness propagate throughout the vault, and the other is anchoring cables or rods to the lamellas, in the two main directions laying diagonally with respect to the lamellas – which are, the longitudinal and transversal directions relative to the vaults.

The enclosure of these vaults can also be built in more than one way. One way is similar to the first bracing one we have just seen: it is based on enclosing the space between the lamellas with rhomboidal opaque panels. The other way is by laying a secondary frame of purlins (Figure 1.43) – and, possibly, a tertiary frame of battens transversal to it – on top of the lamellas, so as to make them work as intermediators between the lamella structure and the transparent enclosure. The added benefit of the latter solution is that it allows the logic of the transparent enclosure to change from one based on rhomboids to another based on parallelepipeds. In the case in which the transparent enclosures are made of corrugated plastic sheets, which have a certain resistance to bending of their own, large-spaced purlins without transversal battens are usually sufficient as an intermediate structure. Another type of enclosure that is well-suited to the presence of purlins is that of plastic films.

When the bays between the lamellas are closed by panels, a major technical challenge is how to make the joints between panels waterproof, especially at the top of the vault, due to its little or absent slope. This can be obtained by recurring to a curtain-wall-like configuration based on pressure caps (Figure 1.7, Volume 3), or with sealed butt-joints between panels (Figure 1.9, Volume 3).

The fact that in the paneled enclosing scheme, the bays, and therefore the panels, in lamella vaults, are not rectangular, the transparent panels usually are not constituted by sealed-perimeter double glazing, except in high-end applications, because sealed-perimeter double glazing is costly when the shape of the panel is not parallelepiped.

1.9.2. Geodesic domes

Geodesic domes are framed lattice-like structures having in common with lamella roofs the fact that they derive from the repetitions of a simple polygonal shape, like the triangle, and the fact that each part of the structure (except, of course, the foundation) is executed in the same way as any other part. But, differently from lamella vaults, the connections in geodesic domes (at least in the most basic configurations) are located at the end of each timber, of each shaft. What is obtained is a trussed structure that is shape-resistant as a result of its spatial layout, of its tridimensional curvature, which, at a structural level, allows it to work as a membrane. This entails that the various parts of the dome work either in tension or in compression, tangentially to the dome itself, but never in bending (except as regards the bending forces produced by the tendency to buckle). Ultimately, this entails that the connections in these domes must be capable of working both in tension and compression, but they do not need to be bending-moment-resistant. Structurally speaking, they can be hinges.

A solution for creating these connections, in the case of geodesic domes based on triangles, is preparing the end of each shaft, of each timber, giving it a shape that makes it possible to execute the butt-joint between the three shafts at each node easily (Figure 1.44(a)). A way to obtain this is by executing the joints with cover plates suitable for resisting both tension and compression (Figure 1.44(b)).

The hardest problems to solve for enclosing geodesic domes are of the same type as those encountered for lamella roofs – just bigger. Like in lamella roofs, the main two solutions are: (a) basing the enclosures on the use of panels (Figure 1.44(c)) – in which case, the main problem is how to obtain water-tightness and vapor-tightness at the joints between panels; and like in lamella roofs, this problem can be addressed by pressure-capping the joints, or by butt-joining and sealing them; or (b) recurring to secondary intermediate structures, like purlins, between the domed frame and the enclosure. The construction of the enclosures, in any case, is complicated by the fact that the most common solution of standard corrugated sheets is not fit to the task due to the double curvature of the domed shape. This is particularly true for small-sized domes. Even the usage of plain transparent films is problematic for small-sized domes.

A specific case of geodesic domes enclosed with an industry-standard panel-based solution is constituted by the inflated ETFE cushion-enclosed geodesic domes, of the kind shown in Figure 3.10, Volume 3.

Figure 1.44(a). *Example of a geodesic dome constructed with timbers in El Paso, Spain (artistic concept by Henry Nold). In this case, the timbers at the connections are cut to shape so as to be joinable with butt joints. Photo by Originalausgabe, 2021, Creative Commons License*

Figure 1.44(b). *Example of a geodesign dome built with connections using cover plates, suitable for simplifying the construction process. Photo: Shikha Jhaldiyal, 2019, Creative Commons License*

Light Frames (Wooden Frames) 59

Figure 1.44(c). *Example of a geodesic dome enclosed with opaque panels cut to shape and waterproofed at the joints. Photo: Ecosolardome, 2012, Creative Commons License*

2
Timber Frames

Timber-frame structures are characterized by widely spaced columns or posts rather than tightly spaced studs/mullions, and by widely spaced beams rather than tightly spaced rafters or joists as primary members. As both the structural elements and the connections used in timber frames are less numerous than in light frames, the connections bear greater forces and therefore need to be stronger than those in light frames. This makes them more difficult to execute. Another difference between timber frames and light frames is that, in the former, the structural elements are usually spaced farther apart. Plywood panels (OSB panels, masonite panels or shear-resistant panels in general) are not large enough to brace the structural bays, and therefore the bracing function must be performed by another system that is suitable for sustaining concentrated stresses. The result is that bracing a timber-frame structure is usually more difficult than bracing a light frame. On the other hand, timber frames make it possible to do things that would be very difficult with light frames, like creating daring cantilevers or tall structures.

The distinction between timber frames and light frames is more dependent on the spacing of the column and beams than their sturdiness. Indeed, although the simplest timber-frame schemes are characterized by massive and whole-span structural elements, many kinds of timber frames can be constructed with small wooden elements characteristic of light frames. This is the case for light-frame trusses or light-frame portals arranged with interaxes typical of timber frames (see Figures 2.77 and 2.84). The construction elements that the parallel frames are made of are of the light-frame type, but the frames assembled from them in composite or trussed configurations are suitable for playing a role analogue to their sturdier and simpler timber frame counterparts.

For a color version of all figures in this book, see www.iste.co.uk/brunetti/greenhouses2.zip.

As seen, timber-frame greenhouses (including those built with a small section of timber) are usually more complex than light-frame ones, because they usually derive from a combination of timber-frame orders and light-frame orders. In this arrangement, when the timber-frame part is conceived to take care of the general structural requirements of the greenhouse, it may sometimes end up being conceived in a bespoke manner; on the other hand, that bespoke timber-frame part allows the light-frame parts to be conceived in an ordinary, non-bespoke manner (see section 2.1.2).

On references

For a book on modern timber-frame construction criteria suitable for self-building, see Roy (2004). For books about modern timber-frame construction inspired by traditional principles, see Sobon and Schroeder (1984), Benson (1997) and Chappel (2011). The already cited books by Feirer and Hutchkins (1986), Canadian Wood Council (1991) and Spence (1999) are also suitable for timber-frame construction.

The books by Borer and Harris (1994–1997), and the last section of the book by Eccli (1976) are interesting references for the Segal method – a low-cost, efficient timber construction method using light portal frames, that can also be used for greenhouse framing.

2.1. Intermixing light-frame parts into timber frames

Timber frames and light frames are almost unavoidably intermixed in timber-frame greenhouses, unless the greenhouses are enclosed with transparent film and the films are installed with no secondary structure (this is possible when the timber frame is arranged in such a way that light frames are not required to instal the films). Another situation in which a timber frame could work alone is when its structural elements are so tightly spaced that they can support the transparent panels directly, without the aid of light-framed elements.

The fact that timber frames and light frames are intermixed most of the time also implies that one of the two frame types (or both) can be substituted with an equivalent one made with other materials; steel, concrete or masonry, in the case of timber frames, and steel, aluminum or PVC, in the case of light frames. As it has previously been seen, wooden light frames can easily be substituted by light-gauge steel frames; the same goes for aluminum frames. However, both light-gauge steel frames and aluminum frames have thermal disadvantages with respect to wooden frames, unless they incorporate thermal breaks, which, on the other hand, increase the complexity of the configurations involved.

Timber frames and light frames can principally be intermixed in the following ways.

2.1.1. *Light frame completely additional to the timber frame*

The most straightforward way of combining a timber frame with a light frame is to make the two coexist so that each one maintains its most favorable arrangement. In the collaboration scheme, the timber frame works as the principal load-bearing structure and the light frame works as the secondary structure bearing the enclosures.

The part of the structure that requires bracing is the timber frame, while the light-frame system does usually not require any bracing, except in the cases in which the spans between timber framing elements are unusually large. This is because, as the light frame is anchored to the timber frame, if the timber frame stands steadily, the light frame will as well.

The two main solutions of this type are as follows.

2.1.1.1. *Light-frame mullions/studs anchored to the beams*

The usual arrangement is when the principal structural elements of the light frame (studs/mullions or transoms/purlins) are parallel to the principal ones of the timber frame – which, in the case of the façades, are the posts, and, in the case of the roofs, are the sloped beams (rafters). In this scheme, the secondary elements of the timber frames are the transversal beams, or the joists, and the secondary elements of the platform frame are the purlins.

In the arrangement, the timber-framed posts bear the timber-framed beams, which in turn bear the light-framed studs/mullions, which finally bear the light-framed purlins/joists (see Figure 2.4). Also, although the dimensional modules of the two structural orders usually are the same (i.e. the interaxis between mullions/studs is commonly a fraction of that between posts, and the interaxis between rafters or purlins is commonly a fraction of that between beams), they could potentially be different (see Figure 2.4, bottom row). This happens, for example, when the columns of the timber frame are outside of the modular grid of the light frame, or when the beams of the timber frame are outside of the modular grid of the transoms or purlins.

However, this construction solution potentially allows the modular grid of the mullions/studs to be completely independent of that of posts, and the modular grid of the rafters or purlins to be completely independent of that of the beams. It even allows the structural grid of both the mullions/studs and the rafters or purlins to be independent of that of the columns (or posts) and beams.

2.1.1.2. *Light-frame purlins anchored to the columns/posts*

The alternative arrangement entails using the purlins as principal elements of the light-frame system, and anchoring them to the posts (and beams) of the timber frame. In those cases, there is no alternative to using purlins rather than transoms as horizontal elements of the light frame, because the elements in question need to be capable of spanning fully from column to column[1]. The main feature of the described structural scheme is that it makes the free span between the purlins considerable, which can be addressed by making the purlins sufficiently strong to sustain the structural task, which in turn may be obtained: (a) by sizing the section of the purlins so as to make them sturdy, to a degree that is more typical of timber-frame components, and (b) by keeping their section lean, but tall, to a degree that is more typical of rafters or joists.

It is also interesting to note that this scheme, as in the previous one, allows the possibility of disconnecting the structural grid of the beams from that of the purlins, the structural grid of the posts from that of the mullions/studs or both structural grids simultaneously.

2.1.2. *Combining timber frames and light frames*

The usual way of combining timber-frame elements and light-frame ones is that of keeping the two domains distinct (see Figure 2.1). The implementation of this strategy requires the investment in a greater amount of material than needed by intermixing the two orders (see Figure 1.39), but it is simpler both at the design and construction level. As the enclosures of timber-framed greenhouses are developed with light-frame techniques most of the time, the solutions that have been described for the construction of envelopes in light frames also apply to timber frames.

---

1 Indeed, the difference between purlins and transoms is that the purlins are continuous along the studs/mullions/rafters, while the transoms interrupt themselves when they encounter the studs/mullions/rafters.

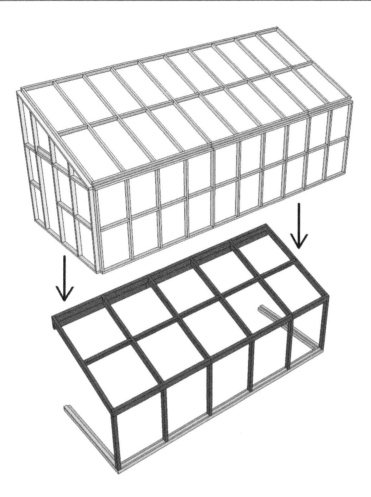

Figure 2.1. *Combining a timber-frame structure with a light-frame envelope may be seen, conceptually, as laying out the latter on the former*

NOTES ON FIGURE 2.1.– It should, however, be noted that the scheme shown here does not involve the construction sequence, because the light frame is assembled piece by piece on top of the timber frame, not translated rigidly onto it. It should also be noted that the strength of the light-frame components can be reduced with respect to that required for light-frame-only greenhouses, as allowed by the support of the timber frame. The timber-frame elements in the image shown here and in the following have been given in a red color for the sake of clarity.

Figure 2.2. *A greenhouse between two buildings in lower Bavaria, Germany. Example of a timber-frame order combined with a light-frame one on top of it, forming the curtain-wall configuration holding the glass panels. Photo: Burkhard Mücke, 2018, Creative Commons License*

Timber Frames 67

Figure 2.3. *How the light-frame components are arranged in relation to the timber-frame structure depends on which structural members are used to anchor the light frame to the timber frame. When the vertical or sloped elements of the timber frame are used, the primary structural elements of the light frame are the horizontal ones. When the horizontal elements of the timber frames are used, the primary elements of the light frames are the vertical and/or sloped ones – like in this example*

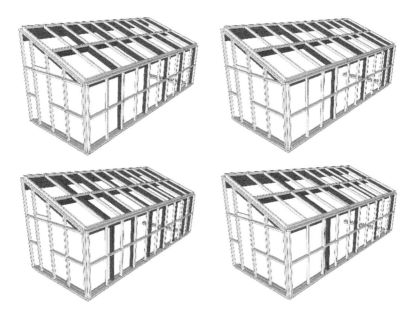

Figure 2.4. *In these arrangements, the light-frame structure of the roof is similar to that of the walls, but secondary cross-beams have been added at mid-span of the timber-frame roof to support the light-frame rafters*

NOTES ON FIGURE 2.4.– In the four images, it is also shown that the spacing of the timber-frame structure can be independent to that of its light-frame complements. In these examples, the module of the timber frame is indeed, respectively (from left to right and top to bottom), equal to two times the module of the light frame, then three, one and a half, and two and a half. Of course, the different spans at play will influence the choice of the size of the beams and transoms, as well as the size of the columns and mullions.

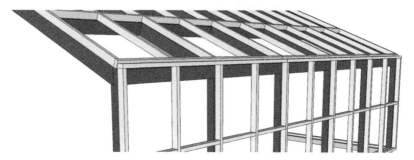

Figure 2.5. *In this variant, the upper edge of the light-frame wall is higher than that of the timber frame. This opens up the opportunity to increase the extension of the glazed front wall to its maximum limit in height, with respect to the extension of the roof*

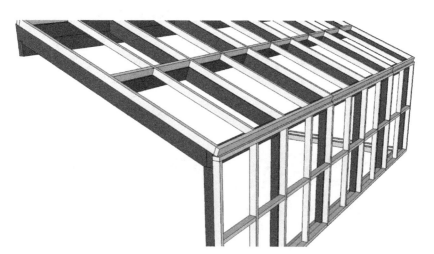

Figure 2.6. *In this variant, the light frames of the roof and the wall are wider than the timber frame. This opens up the possibility of increasing the glazed extension of the roof at the expense of the height of the glazed wall*

Timber Frames 69

Figure 2.7. In this variant, the front wall is cantilevering out from the post to make the plane of the side wall flush with the side of the front wall. This can be obtained by making the last structural bay of the light-framed wall larger or making the distance between the end posts smaller. In the former case, glazing of a slightly larger size would be needed. In the latter case, the same glazing size as the other bays would be required

Figure 2.8. In this variant, (a) the height of the light frame of the front wall can reach its possible maximum, so as to maximize the solar gains from the wall, and an edge light-frame fascia beam (header joist) has been added on top of the timber-frame edge beam, so as to increase the strength of the supports; (b) the side walls have been made longer, so as to be flush with the front wall. The width of the glazing of the side façade is increased at the expense of the glazing of the front façade

Figure 2.9. *In this variant, the section of the light-frame fascia beam (eave beam, header joist) has been made upright to increase the extent of the glazing on the roof. A gutter may be positioned in front of the fascia beam*

Figure 2.10. *A hybrid solution is shown, in which the light-frame rafters sit on top of the light-frame mullions with the mediation of a double plate, which adds rigidity to the assembly*

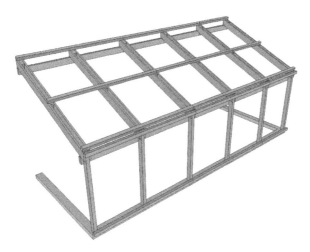

Figure 2.11. *Purlins are added onto this timber frame, both on the roof and on the front façade (at the top and bottom of it). These purlins may be the primary elements of the glazed roof (unless, at least, it is enclosed with transparent films), combined with secondary mullions/rafters/studs. In that case, the glazed roof would be constituted by glazed panels laid horizontally. As an alternative, these purlins may support a light-frame system of mullions as primary elements (which would, in that case, be constituted by vertical or sloped primary elements) and transoms as secondary elements*

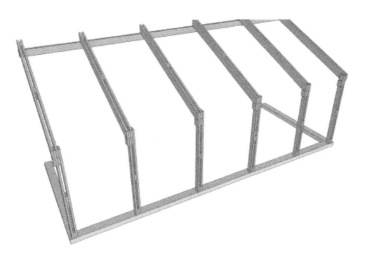

Figure 2.12. *These half-frames are realized with double posts and double beams built with light-frame-sized lumbers. The joint between them is lapped and asymmetric, and the double elements have blocks positioned between them to increase their resistance to lateral bending and therefore buckling*

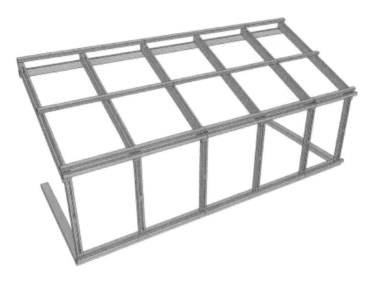

Figure 2.13. *This timber-frame variant constitutes the completion of the previous one with purlins*

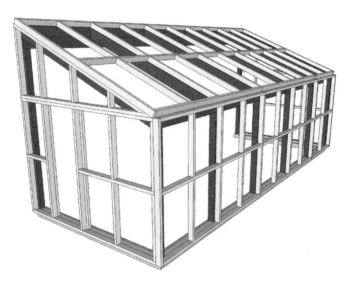

Figure 2.14. *In this timber-frame and light-frame combination, the primary elements of the light frame are the horizontal ones. Indeed, they are continuous and work like purlins. The mullions/studs/rafters get interrupted at each connection by a horizontal component. This avoids requiring timber-frame mid-span beams on the roof*

Timber Frames 73

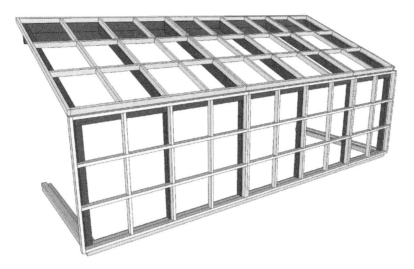

Figure 2.15. *In this variant, a shorter (with respect to the previous image) interaxis between purlins/transoms has been chosen for the roof and the wall*

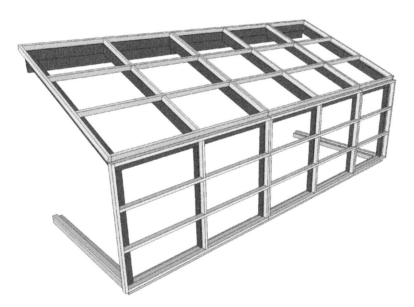

Figure 2.16. *In this variant, the transoms have been more clearly transformed into purlins, and their section therefore needs to be increased, especially if glass panels (heavier) are used for glazing rather than plastic panels (lighter)*

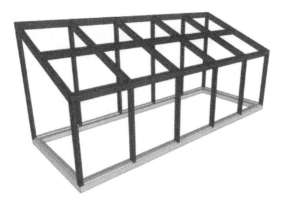

Figure 2.17. *An example of a timber-frame structure for self-standing solar greenhouses, at its maximum simplicity*

NOTES ON FIGURE 2.17.– The massive back wall is not a part of the load-bearing structure. Therefore, it could be absent, or it may be external to the timber-frame posts, or internal to them – that is, standing within the greenhouse. If the massive wall is absent, the envelope at the back façade could be made opaque and insulated to improve solar gains or could be made transparent. But even in this case, the absence of the thermal mass at the wall would create wider daily thermal swings indoors: higher temperatures during the day and lower temperatures during the night: conditions typically suitable for a greenhouse used as an air collector (e.g. for pre-heating the air for a building, possibly in combination with forced air circulation), but less suitable for growing plants.

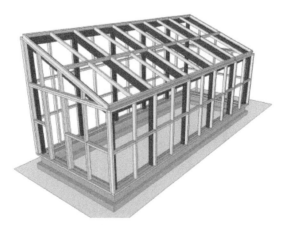

Figure 2.18. *The back wall and the front wall of a solar greenhouse may be built in the same way, at the expense of the thermal mass*

NOTES ON FIGURE 2.18.– If this is done, however, to make the greenhouse "solar", suitable for passive solar gain, at least the back wall (the wall facing away from the equator, oriented to the north in the Northern Hemisphere) should be made opaque towards the inside, in order to allow for solar gains, preventing the incoming radiation from exit without being gained. The simplest way to ensure this is by positioning non-low emittance (i.e. non-shiny-reflective) shading devices, like canvases, cloth, insulated blankets (rollable, for example) or even Venetian blinds, on the inner back wall. The addition of thermal masses, for example, as closed water barrels, in these cases would be beneficial.

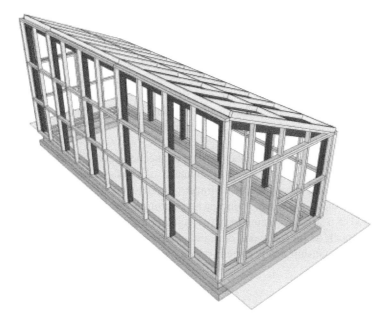

Figure 2.19. *The construction at the ridge can be similar to that at the eave*

2.2. Connections in timber-frame greenhouses

Modern timber-frame connections constitute a substantial evolution from traditional ones. Traditional timber-frame connections, the simplest of which are the tenon-and-mortice joints, were characterized by tight-fit collaborations and locked geometries, while modern connections are usually characterized by the use of auxiliary connectors and devices (mostly made of rolled steel) that simplify the woodwork and reduce the likelihood of excessive concentrated stresses in the wood matter.

The tenon is the convex, protruding part of the connection, and the mortice is the concave, hollow one. The tenon-and-mortice connection requires the components to be joined (beams, columns) to be worked out, so as to create mortices into them, in order to make them capable of hosting the tenons of the other component to be joined. In today's connections, when connections are hollowed, it is usually with narrower and deeper cuts, suitable for hosting internal steel plates in them. Today's strategy reduces the subtraction of material in the structural elements and increases the strength of the connections.

Traditional timber connections, of which the tenon-and-mortice connections are by far the most important and common, have not been abandoned, but tend to be used less and less. Some versions of the tenon-and-mortice joint are still in use today, both at the craftsman's level, and in modern furniture and house-building, where the connection scheme has been radically simplified thanks to the radical simplification of the shapes that the timbers are cut into.

Two niches in which tenon-and-mortice connections are still thriving are North America, where there is a strong community of builders devoted to traditional timber framing, and Japan, where a climax of efficiency and complexity in locked wood joints (Engel 1987) has been reached during a broad span of centuries. This has resulted in a strong contemporary heritage in modern Japanese carpentry, which relies very much on precise wood cuts and precise interlocking of parts and requires a level of workmanship that is currently well beyond the average in the international scenery.

2.2.1. *Traditional connections in timber frames*

The main problem with traditional timber-frame connections is that they require a great deal of precision, workmanship and judgment skills to choose and execute the right wood cuts. Precision in traditional carpentry was slowly obtained through gradual refinement, but modern machines are much quicker and efficient than humans for obtaining precision. However, ordinary machines cannot easily perform all the complicated composite tasks that human craftsmen can perform yet, nor are they as good as humans in understanding if the grain of the timber in the zone that has to be worked out is suitable for undergoing that modification and performing well once fit into its place in the connection itself. The consequence of these limitations is often a simplification of the details in machine-made connections, lower quality when they try to mock traditional ones, and potentially some risk of underperformance. As a result, although many complicated connections from tradition – like dovetail connections, miter joints and finger joints – still survive in modern carpentry, they tend to be more and more superseded by modern connections, aided by auxiliary steel components.

2.2.2. Modern connections in timber frames

As seen, the timber part of modern timber connections is simpler than that of traditional ones and often relies on the aid of steel (or sometimes aluminum) connectors: heavy-gauge or light-gauge (in timber frames, more often the former, in light frames, more often the latter) auxiliary connectors, like plates, angles, together with additional connectors like bolts, screws or nails. The principal functions of auxiliary connectors in modern timber connections are: (a) holding the connections together and (b) transmitting the forces across the joined elements. The two functions can be decoupled or be performed together. In Figure 2.45, for example, the function of holding together the connection is performed by screws, while the function of carrying the loads is performed by the corbel. In Figure 2.34, the two functions are instead performed by the same component, the bolted plate embedded in the joint.

2.2.2.1. Tenon-and-mortice connections

Tenon-and-mortice joints ensure that the quantity of material available in the connected components at the zone of the connection is reduced, in order to make room for the concurrent presence of the other component(s) of the connection. The problem with this fact is that it makes the components to be connected weaker (by subtraction of material), especially in the parts that require more resistance. This is indeed relevant because the resistance of a structure is determined by its weakest parts, the parts that would be the first to fail – not its strongest ones. The connection, in other words, is usually the worst point where a structure can be weakened; but in a tenon-and-mortice joint, this occurrence cannot be avoided.

The geometry of the tenon and mortice connection makes it resistant to shear, but not to bending. Furthermore, a simple and ordinary tenon-and-mortice joint, when loaded with axial forces parallel to the tenon, can only resist compression, but not tension; unless it is of the dovetail type (see Figure 2.20): that is, an open tenon-and-mortice joint, in which the tenon is larger at it "head" than at its "root". But even in the case of dovetail tenons, the resistance to tension is inevitably weaker than in compression. Furthermore, dovetail connections are only possible (for obvious geometrical reasons related to the possibility of insertion) with bidimensional, "open" tenon-and-mortice joints (of the kind shown in Figure 2.21), and are not possible in tridimensional, "housed" tenon-and-mortice joints (of the kind shown in Figure 2.25).

Figure 2.20. *Example of multiple dovetail connection*

The most specific difficulties to be overcome in working out a tenon-and-mortice joint, even in the simplest configurations, are (a) producing a strong tenon, taking into account the local nature of the wood grain, so as to avoid it becoming weakened by the local irregularities of the wood grain, which may be difficult to detect through a purely visual exam before the timber is worked out, and (b) producing the mortice, especially when the tenon is of the "housed" type, due to the difficulty of giving sharp edges to the cavity in the timber.

The result of these issues is that in modern times dovetail connections are mainly used in furniture production, where a massive recourse to auxiliary connectors would be aesthetically impractical, rather than modern carpentry.

2.2.2.1.1. Open tenon-and-mortice joints

The most modern version of the tenon-and-mortice joint is the open one, thanks to its simplicity. Its simplicity is also the reason why it is still in use today. In the simplest, but weakest version of the open tenon-and-mortice joint, the thickness of the tenon is the same as that of the two sides of the mortice (Figure 2.23), and in the most efficient version, the thickness of the tenon is the same as the sum of the thickness of the two sides of the mortice, in order to make the quantity of material involved in the two connected elements the same. However, the latter solution can only be adopted when the two sides of the mortice are not too thin to be unpractical for fabrication and use – which only happens in large-sized components.

The open tenon-and-mortice joint, statically speaking, is a hinge (more specifically, semi-rigid) connection that is not resistant to tension, but it can also resist shear only in the direction orthogonal to the sides of the open mortice. Consequently, it is rarely used as it is in construction: it is usually combined with solutions for adding shear resistance in the direction parallel to the sides of the open mortice. The most common solution to obtain this is to lock the tenon in place with pegs or bolts (see Figure 2.23) (or, less commonly, nails or screws), at the cost of reducing the resistant section of both the tenon and the mortice. The pegs or bolts make the open-tenon-and-mortice joint capable of resisting both in shear (along the direction parallel to the sides of the open mortice) and (great advantage!) in tension, axially (along the direction of the tenon: this is because they lock the tenon in place, without requiring dovetailing). Regardless of the material of the pegs, the space the pegs occupy into the components entails a reduction of the resistant section of the components. For this reason, pegged open tenon-and-mortice connections are mainly used when it is acceptable that the resistance to tension of the connections is markedly smaller than their resistance to compression. Indeed, this is something that can be obtained when the volume of pegs into the tenon and into the walls of the mortice is markedly smaller than the volume of holes in the tenon and in the walls of the mortice.

The open-tenon solution can be used both when the side planes of the mortice are vertical and when they are horizontal. Of the two solutions, the former is more interesting for connections between beams and posts/columns, because the height of the monolithic sections of the components is substantially higher than in the other case.

Another connection solution is that of sustaining the component to which the tenon belongs with a corbel integrated into the component (at the surface of it, or notched into it) hosting the mortice. Of course, this solution almost only makes sense when the joint is vertical. The pegged (or bolted) solution and the corbelled solutions can also be combined into one composite solution for greater efficiency.

The pegs can be made of wood (usually, strong hardwood) or steel, and should be forced into the cavities of the wood. The bolts, instead, do not need to be forced into the cavities: their retention capability is assured by the bolts' heads, the nuts and the washers. Bolts are therefore a safer solution than pegs where large fluctuations of moisture content can be expected, because those can loosen the pegs. These fluctuations can be especially wide in agricultural greenhouses where the growth is not year-round, but limited to some periods.

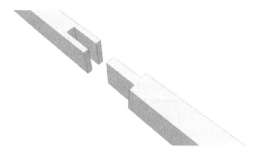

Figure 2.21. *Open tenon-and-mortice joint between two beams*

Figure 2.22. *Open tenon-and-mortice joint between posts and beams*

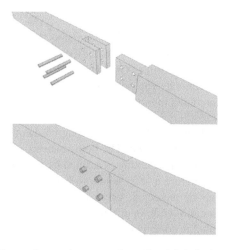

Figure 2.23. *Pegged open tenon-and-mortice joints between two beams*

The orientation of pegged tenon-and-mortice connections is freer than that of non-pegged ones, and this makes it possible to devise them to maximize the connection efficiency. For example, the horizontal open tenon-and-mortice joint in Figure 2.24 does not have to rely on pegs to support shear in the vertical direction, differently from the vertical tenon-and-mortice joint in Figure 2.23. This means that it may need fewer pegs, if the main direction in which it is called to resist shear is vertical.

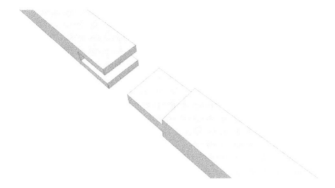

Figure 2.24. *Horizontal open tenon-and-mortice joint*

2.2.2.1.2. Housed tenon-and-mortice

"Housed" (closed) tenon-and-mortice joints (see Figure 2.25) can resist shear in both of the main directions orthogonal to the tenon (the vertical one and the horizontal one), even without the assistance of other connectors. However, they are more challenging to produce than their open counterparts. In particular, what is most difficult to produce is the mortice, because it requires carving out the head of an element in a component. For machines, even CNC[2] machines, what is difficult is carving out the mortice, making its internal edges sharp. Another potential issue regards the quantity of material constituting the mortice in the case in which the volume of the tenon equals the volume of the mass of the mortice "walls". When the section of the elements to be connected is small, the thickness of the "walls" of the mortice can become insufficient for obtaining sufficient mechanical resistance.

---

2 Computer Numerical Control.

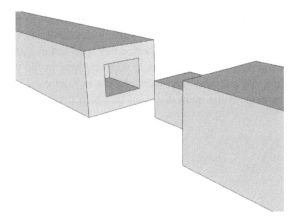

Figure 2.25. *Housed tenon-and-mortice joint. This connection is infrequent between two elements laying along the same axis, but in traditional timber, framing is common between two orthogonal elements (like a beam and a column), or elements laid out diagonally with respect to each other (like a bracing diagonal and a column, or a beam – see Figure 2.26)*

2.2.2.1.2.1. *Housed tenon-and-mortice joint completed with pegs or bolts*

A potential issue of the symmetric housed tenon-and-mortice joint configurations (see Figure 2.25) is that, because the tenon is reduced in two directions, it may not have enough space to host a number and size of pegs or bolts sufficient to assure a satisfactory resistance in tension to the connection. But in construction, the housed tenon-and-mortice joints are usually elongated along one direction (which, for example, in beam-to-post connections is the vertical one), and this situation alleviates the problem of lack of material in the tenon, because it increases the area at the sides of the tenon. This is what happens in the typical tenon-and-mortice joint used between columns or beams and bracing diagonals in North American timber frames, of the kind in Figure 2.26.

The latter connection is a heritage from traditional timber framing. The technique involves embedding the tenon in a mortice in the beam and in a mortice in the column, typically shaped so as to finish the 45° tilt with a further 45° cut. This connection is suitable for working in compression; however, the tenon in it is usually held in place with pegs, in order to give the connection some resistance in tension. But, because in those cases, usually, the housed tenon is of a size that is only sufficient to host one or two pegs at the most, the connection is much more resistant in compression than in tension; which bears consequences on how the bracing arrangement has to be. In these cases, the arrangement of the bracing elements has to be conceived in such a way that all the diagonal forces can be resisted by the bracing diagonals, mainly in compression, which in turn determines the fact that the diagonals must double in number and be placed symmetrically, rather than leaving the choice of making their arrangement symmetrical or asymmetrical. Indeed, placing the bracing diagonals asymmetrically is only possible when the connections and diagonals can resist in both tension and compression (see Figure 2.104).

Figure 2.26. *Structure characterized by housed tenon-and-mortice connections at the diagonal bracings. Photo: Chris Light, 2011, Creative Commons License*

2.2.2.1.3. Mortice-and-mortice joints with added tenon insert

A modern solution (reprised from the past) for obtaining tenon-and-mortice joints simply is constituted by combining two mortices with a housed insert as tenon – in place of the tenon-and-mortice configuration. This solution comes in all possible variants: unpegged and unbolted, pegged or bolted. The solution makes it possible to use inserts of a material that is stronger than that of the mortices, and to make the inserts smaller, with the resistance being equal. However, in order to avoid the possibility of rotation and juggling, the insert should be longer than a tenon rigidly anchored to one end. The internal steel plate connection (see Figure 2.34) can be seen as an extremized derivation from the described solution.

2.2.2.1.4. Tenon-and-mortice connections between beams and columns

In traditional tenon and mortice connections between beams and continuous columns, the column is often partially carved out so as to form a sort of corbel for the head of the beam, at the price of subtracting some more material from the column. When the column is not prepared in such a way, the tenon has to respond to the load of the beam, but the post is spared to be carved out further. This may be especially beneficial for continuous columns – that is, columns that do not stop at the connection. When the column stops at the connection, there is also the possibility of making the beam continuous there, like in Figure 2.26. In those cases, the column (better defined as the "post") can integrate the tenon at its head (or a mortice to be combined with an insert), correspondingly to the fact that the mortice can be located at the downward face of the continuous beam.

2.2.2.2. *Butt connections*

Butt connections share the fact that the "head" of a component is connected to another component: to the end of it (orthogonally or at an angle), to a side of it (again, orthogonally or at an angle) or to the end of it along the same axis.

Unlike butt connections in light frames, butt connections in timber frames, when executed with nails or screws without the aid of plate connectors or angle connectors, are very often too weak to be advantageous. Concentrating forces is what timber frames are about, as much as distributing forces is what light frames are about. From this, it can be said that the type of connection employing steel cover plate connectors, or angle plate connectors, is by far the most common in modern timber frames.

2.2.2.2.1. Butt connections with external hot-rolled steel angles and/or plates

The simplest butt connections entail the use of steel angles (in end-to-side butt connections) (see Figure 2.27), steel plates (in end-to-end axial connections) (see Figure 2.28) or angles and plates (in end-to-end perpendicular connections). When angles are involved (see Figure 2.27), the best connection strategy is bolting, because the bolts can be stressed in compression and tension, not only in shear, while the resistance of screws and even more nails in tension is small (unless they are set in diagonal and mutually divergent arrangements, which are not the easiest to execute). In connections that only involve plates, instead – like in axial connections – the bolts, nails or screws, are called to work in shear, and therefore nails and screws are also suitable (see Figure 2.28). However, when, in other butt connections, bolts are used, they are often also used in all of the other connections, for the sake of simplicity.

Timber Frames 85

Figure 2.27. *Scheme of bolted (or screwed, or nailed) connection between column and beam aided by steel angles*

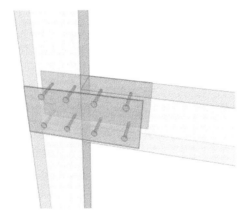

Figure 2.28. *Scheme of bolted (or screwed, or nailed) connection between column and beam aided by external steel cover plates*

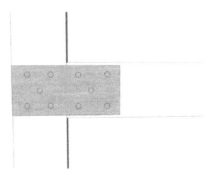

Figure 2.29. *Scheme of bolted (or screwed, or nailed) connection between column and beam aided by external steel cover plates*

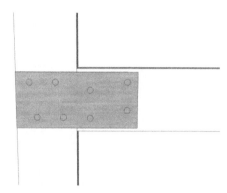

Figure 2.30. *Scheme of bolted (or screwed, or nailed) connection between column and beam aided by external steel cover plates. Here, the bolts are kept un-aligned with the wood grain to decrease the likelihood of splits due to the stresses. This has entailed un-aligning them horizontally with respect to the beam (due to the fact that the wooden fibers are horizontal there) and vertically with respect to the column (due to the fact that the wooden fibers are vertical there)*

Figure 2.31. *Scheme of bolted (or screwed, or nailed) butt connection realized with wooden cover plates (also common with steel cover plates). Due to the shortness of the cover plates, this connection cannot be considered bending-moment-resistant. It is a semi-rigid connection*

Figure 2.32. *Scheme of bolted (or screwed, or nailed) butt connection realized with wooden cover plates (also common with steel cover plates). Due to the length of the cover plates used, this connection attained a non-negligible bending moment resistance*

2.2.2.2.2. Butt connections with tooth plates

External cover plates can also be of the so-called tooth plate (also called toothed plate, or gusset plate) type, light-gauge, as an alternative to heavy-gauge non-toothed plates. Tooth plates are usually rectangular, but many shapes exist in the market. Tooth plates can be used both in light-frame and timber-frame butt connections, but their usage in the latter case is less common, because of the high stresses involved.

Figure 2.33. *Lap- and butt-joints obtained through tooth plate connectors in a timber trussed roof structure*

2.2.2.2.3. Butt connections with internal plate connectors

Steel plates – in this case, heavy-gauge – can also be used as internal plate connectors, (a) of standard type (rectangular, L-shaped, cross-shaped, Y-shaped, X-shaped, etc.) (see Figures 2.34, 2.35 and 2.36) or (b) bespoke, obtained as composite components via welding. Because the plates have to be housed in lodgments into the timber elements to be connected, the solution in question can be used when the remaining thickness of a timber element is not too small – better if not smaller than about 4 centimeters. The described solution is almost only used with bolts (rather than screws or nails), and may sometimes include ring connectors (see Figures 2.39, 2.40 and 2.41). Compared with the solutions based on external plates, those based on internal plates use less steel, and the plates are better protected from fire (being embedded in wood) and more hidden from view.

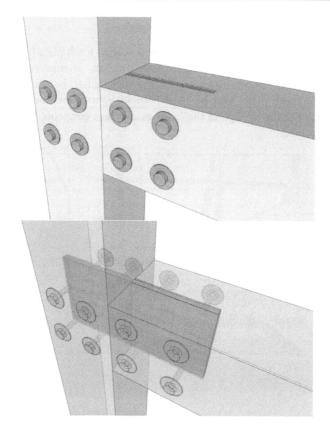

Figure 2.34. *Bolted connection between column and beam obtained via an internal steel plate*

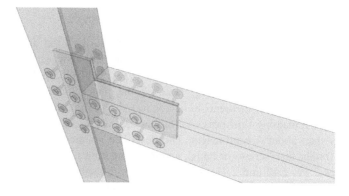

Figure 2.35. *Heavy-duty bolted connection between column and beam obtained via a T-shaped internal steel plate*

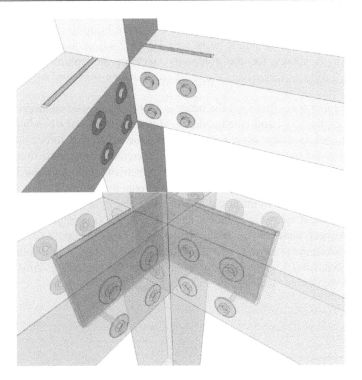

Figure 2.36. *Bolted connection between a column and two beams obtained via an internal, composite steel connector*

2.2.2.3. *Lapped connections*

Depending on where the overlap between the connected elements takes place, lapped connections can be arranged as end-laps (see Figure 2.38, left), T-laps and intermediate-laps (see Figure 2.38, right). When the elements to be connected are perpendicular to each other rather than parallel, the connection is called edge-lap (see Figure 2.37, left). A disadvantage of lap joints, when they are not executed in cover-plate configurations, is that the axes of the joined elements lie on different planes, which creates secondary bending forces and torque in the elements themselves, in a way that may perturb the local mechanical response of the elements involved and is difficult to predict with precision, even with computer models – which commonly leads to addressing these predictions in a simplified way, by means of safety factors. Another disadvantage of lap joints is that their strength mainly depends on the connectors – bolts, screws or nails (of which bolts are most commonly used due to their greater strength and structural efficiency).

A possibility opened by lap joints in the context of beam-to-column connections is that of the double continuous beam (see Figure 2.44) or the double continuous column, which allows for the continuity of both the column and the beam (see Figure 2.37).

When neither the column nor the beam is double, the lapped connection between the two is asymmetric, and entails additional internal secondary bending moments and torque; but in many situations, both of those additional forces are small in comparison to the primary bending forces, and therefore add up little to the required resistance of the elements and connections. Symmetrical configurations derive from doubling the column or the beam, and the additional secondary stresses at play can be reduced consistently by inserting intermediate blocks at intervals along the free span of the double elements to reduce their free length (see Figure 2.70). Because the ratio between bending resistance and span length is quadratic (see Volume 4, Chapter 2), even the placement of a few blocks along the length of a double column or double beam has a great effect of reduction of the secondary stresses in each segment of the double elements.

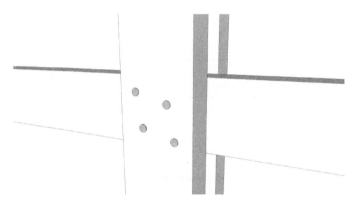

Figure 2.37. *Screwed, nailed or bolted connection between a double column and a beam. The alignment of the screws/nails/bolts has been avoided to decrease the risk of wood splitting along the grain, but may be unnecessary in most cases*

The strongest screwed or nailed lap joints have a symmetric distribution of screws/nails, meaning that approximately the same number of screws/nails enter at the two faces of the connection. However, in this case, there is the risk of the screws/nails being too near to each other (like in Figure 2.38), making the connection weaker.

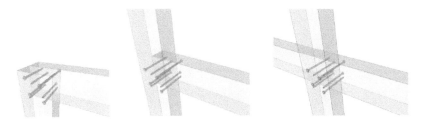

Figure 2.38. *Screwed or nailed connections between column and beam. To increase the resistance, the screws or nails can be inserted from both sides of the elements, as demonstrated here. In these cases, however, care should be taken that the screws/nails do not end up too near to each other. This risk lurks in these examples. As a rule, a distance of a minimum of three diameters should be kept*

2.2.2.3.1. Utilization of shear connectors

In this type of configuration, the lap connections are reinforced by shear connectors (ring connectors) placed at the boundary between the two timber elements in a lap joint, centered on the bolt that they serve, in order to increase the surface of exchange between the bolt and the two joined timber elements. Shear connectors are fundamental to make lap connections stronger and increase the range of the cases in which they can be used.

A ring connector can be: (a) of the solid type (the strongest type, which, however, needs the preparation of a cavity in the two elements to be joined), (b) of the toothed type (see Figure 1.16) (weaker, but directly connected to the bolt, and not requiring any preparation on the wooden elements) or (c) of the hollow ring type (low cost, but needs a lodgment to be prepared in the two timber elements; and weaker than case (b), because it is not directly connected to the bolt, but indirectly connected to it, via the wood).

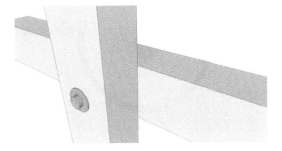

Figure 2.39. *Bolted lap connection between continuous column and continuous beam*

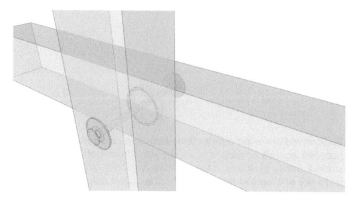

Figure 2.40. Bolted lap connection between continuous column and continuous beam. Here, the ring (split-ring) connector is visible in transparency

Figure 2.41. Types of ring connectors. From the TECO catalogue. Left: tooth plate; center: split-ring; right: shear-plate

Figure 2.42. View of the lodgment of a split-ring ring connector in a damaged timber element. Photo: Hab044, 2013, Creative Commons License

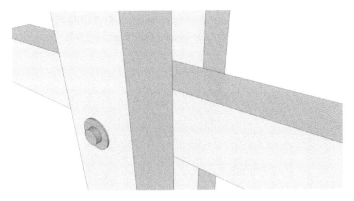

Figure 2.43. *Bolted lap connection between a continuous double column and a continuous beam (see also Volume 4, Figure 1.18)*

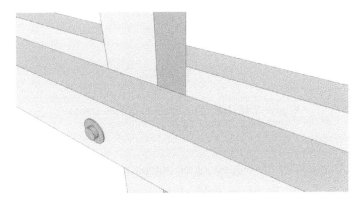

Figure 2.44. *Bolted lap connection between a continuous column and a continuous double beam*

2.2.2.3.2. Utilization of internal corbels

An alternative to increasing the resistance of a lap joint is constituted by adding internal corbels to support the beam in between the two halves of a column (see Figure 2.45). The task of the corbel can be performed by the same type of component performing the function of mid-span-blocks (see Figure 2.12). The corbels can be anchored by bolting (with or without shear connectors) or (in a more distributed fashion, and with greater technical difficulty) by screwing or nailing.

Similarly, when the internal corbel "blocks" (support blocks) can be integrated between half-columns in double-column/single-beam configurations, they can be integrated as external corbels at the two sides of double-beam/single-column configurations, in support of the double beams (see Figure 2.97).

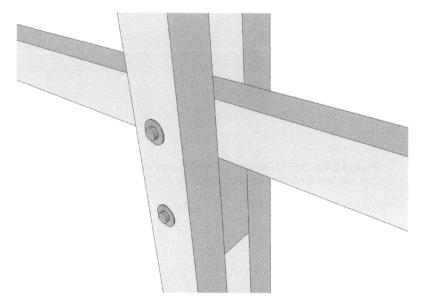

Figure 2.45. *Lap joint between double column and beam supported by an internal corbel*

2.2.2.3.3. Butt-joining the lapped-joined components

Due to limitations of the strengths of the components, in some situations, the necessity arises to butt-join two beams or two columns along the same axis. When that situation regards beams, due to the lack of collaboration between the ends of the joint, and due to the difficulty of adding extra bolts/screws/nails to them, the load-bearing capacity of the connection may be strengthened by means of cantilevering corbels or cover-plates. The possibilities are analyzed in the following sections.

2.2.2.3.3.1. *Assuring continuity between double continuous elements with cover plates*

As said, the continuity between double continuous elements (double beams or double columns) can be assured by using cover plates to execute the butt joints. Ideally, those connections should be placed in the zone at about 1/3 of the span between supports in continuous beams. For a more exact view, they should be positioned at the expected inflexion points in a continuous beam – that is, the point at which the bending curvature changes its sign and direction. This is because, in the places with no curvature, the bending moments are 0, and a hinge – or, in any case, a point weaker in bending than the others, a semi-rigid joint – would have its most appropriate location there. This is also the principle of Gerber beams[3].

An alternative to solving the connection in question with lapped auxiliary elements like cover plates is to make one side of the connected elements (beams, typically) single, and the other side double (see Figure 2.46). The connection between the single-structural element and the double ones parallel to it, in that case, may resemble the other connections used for joining a single element to the double elements orthogonal to it (in the case of a single column connected to a double beam and of a single beam connected to a double column).

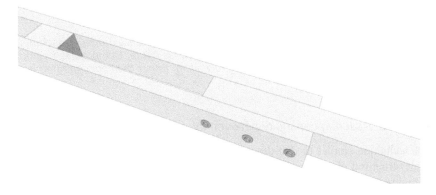

Figure 2.46. *Lap joint between a single beam and a double beam*

An advantage of the kind of solutions in question is constituted by making it possible to build frames using full-length and continuous structural components despite using shorter-length components. Another advantage, deriving from the compensation taking place because the elements are paired, is that the solution is also very well executable with ordinary sawn lumbers; that is, it does not necessarily need glue-laminated lumber – a great thing for sustainability, because synthetic glues are not the healthiest thing on earth. The technique as a whole can indeed be used as a lower-cost alternative to glue-laminated lumber.

---

3 For more information about Gerber beams, see: Lakshmi M. Vemuri and Alfredo E. Bustamante (2018). Updated PROVISIONS and standards ensure safety of cantilever girder structures. *Civil + Structural Engineer Magazine* [Online]. Available at: https://csengineermag.com/gerber-girders-from-forensic-investigation-to-repairs.

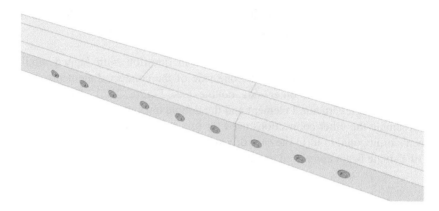

Figure 2.47. *Composite beam obtained by bolting. As an extreme consequences, this solution brings the scheme shown in the previous figure. Compositive beams can also be obtained by distributed screwing or nailing, rather than concentrated bolting, with similar overlapping criteria*

2.2.2.3.3.2. Assuring continuity between lap-joined single elements

A technique similar to that described for joining the components of continuous double-structural elements (columns or beams) or composite elements can be used for joining single-structural elements. The solution can be symmetric or asymmetric. In the symmetric solutions, two elements may be butt-joined and overlapped to two auxiliary segments of the same type, for a length of about 1/3 of the structural span into which the connection has to operate, to which it will be lap-connected (obtaining a connection configuration of the kind shown in Figure 2.32). In the asymmetrical solutions, the two timbers to be joined are instead lapped to each other – again, for a length of about 1/3 of the free span – laying side by side along parallel axes. In the latter configuration, however, the elements will be involved in rather substantial additional torque, which results in the necessity of a certain oversizing of the elements themselves. This can only be had in moderately stressed configurations, not in heavy-duty ones. The other possible asymmetric solution, which is that of using a single cover plate (i.e. a cover plate at only one side of the joint) is, instead, most often, not strong enough to be of any practical use.

In any case, the joint should be located at a point of inflexion (i.e. where there is an inversion of the sign of bending moments, of the curvature under load), in order to minimize the bending stresses in the zone of the joint. The 1/3 criterion is aimed at that, but is an approximated measure. The inflection points can be more finely specified, via detailed structural analysis, depending on the geometries at play.

2.2.2.3.3.3. Assuring continuity between mutually orthogonal pairs of lap-joined timbers: the lapped asymmetric double-column/double-beam frame solution

The technique of lapping pairs of elements can also be applied so that both the columns and the beams are double and made rigid, with intermediate blocks placed at intervals of 1/3 of the spans. The intermediate blocks between double elements (especially columns, with respect to which the risk of bucking is particularly relevant) may be advantageous not only at a structural level, but also for fire protection and aesthetics. The described solution is the ultimate, strongest and most general application of the concept underlying the technique of lap-joining double columns or beams (see Figure 2.68), but is still seldom used in practice.

A clear disadvantage of this solution is that the central axis of the double beams is off-center with respect to the central axis of the columns, which produces additional bending moments in the columns, and, secondarily, torque in the beams, both deriving from the fact that the two elements forming a column are loaded differently. Theoretically, the element that is central with respect to the beam bears about 3/4 of the load, and the peripheral one bears 1/4 of it. In practice, the lateral rigidity of the joint and the presence of the intermediate blocks tend to reduce this difference in loading. The additional bending actions, torque and asymmetrical loading mean that some redundancy should be assumed in the structural calculations (e.g. via safety factors), both for the columns and the beams.

A structure built in the described way along the principal beam direction may use beams located in the same plane as the column, or beams located on one side of the plate of the columns, and/or may be constituted by continuous double beams made rigid by intermediate blocks (see Figure 2.46, left) located at intervals of about 1/3 of the span (like in Figure 2.69).

2.2.2.3.3.4. Pursuing functionality in the connection between double beams and secondary beams and/or joist

The solution of double beams is particularly advantageous when secondary beams sit on top of the double beams, because, in this way, the secondary beams load the two underlying beam components uniformly (see Figure 2.70), even when they are at the edge position in a frame (see Figure 2.96). There is a possibility of anchoring the secondary beams on the same side as the primary ones, asymmetrically, but this means that the external components of the double primary beams receive the force due to this load asymmetrically and less efficiently, with torque as an effect. This, in turn, implies the necessity to oversize the involved structural elements. The best way to reduce the disadvantageous effect of torque is to use intermediate blocks at each location in which the double primary beams meet a secondary one at the side (like in Figure 2.100).

2.2.2.3.3.5. *Creating intermediate floors between mutually orthogonal lapped-joined elements*

When the space of a greenhouse is partially occupied, at an intermediate height level, with a floor – this can happen, for example, due to the presence of a balcony – the suspended floor can be built using secondary beams in greater number, in order to make them work as joists (see Figure 2.94), or, as an alternative, joists can be added to the secondary beams (see Figure 2.99).

2.2.2.4. *Lapped half-wood connections*

Lapped half-wood connections can be configured – like simple lap connection – as end-laps (see Figure 2.48, right), T-laps and cross-laps (see Figure 2.48, left). The difference between half-wood connections and ordinary lap connections is that in the former, the thickness of the elements at the connection is cut in half, in order to make it possible for the elements to fit at the connection without increasing the overall thickness at the joint beyond that of a single element (see Figure 2.44). Half-wood connections have the advantage that the axes of the elements involved are kept on the same plane – or, in any case, on planes that are nearer to each other than they would be otherwise – and have the disadvantage that, being the section of the timbers cut in half at the joint, the connection becomes, by design, the weakest point of the components. For this reason, this kind of connection is seldom used in modern carpentry. The situations in which it is most often used is in the timber plates against the foundation walls (see Figure 2.48, right), because that timber is not stressed during bending. Other general situations of use are those in which the connections are not at the center of the timbers, but rather at the ends, and the bending stresses are small.

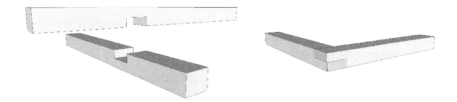

Figure 2.48. *Left: exploded view of a crossed half-wood lap joint. Right: end half-wood lap joint, as it is sometimes used in thick sill plates on top of foundation walls*

2.2.2.5. *Scarf joints*

Scarf joints are diagonal joints for obtaining bending-resistant butt connections, without relying on bending-resistant structural aids (indeed, they are mostly pre-modern), and are especially useful in combination with gluing, because, with respect to simple butt joints, they increase the surface of contact between the end of the components to be joined (see Figure 2.50). For this reason, together with finger joints, scarf joints are suitable for use in the connections internal to the glued-laminated lumber components. In the context of joints between sawn timber components, the type of scarf joint that is sometimes used is the one with key locks (addition of keys that block the timbers in position – see Figure 2.49): a solution deriving from the ingenuity and inventiveness of traditional builders.

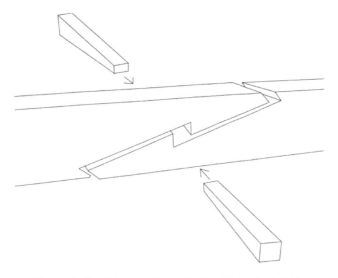

Figure 2.49. *Scheme of scarf joint with keylock. This traditional configuration produces a certain bending resistance*

Scarf joints can also be executed without gluing, in which case they entirely rely on the key lock to give bending resistance to the assembly. In all cases (with or without keylocks), scarf joints are only used in beams and not in columns, because their resistance to buckling would be low.

The reason why the solution of scarf joints with key locks is not common in modern carpentry is the technical difficulty of execution, and, consequently, the time required to construct one. Today, scarf joints are superseded by gluing in finger-joint configurations, which in turn are reserved for factory-made situations, and are tendentially avoided on-site.

Figure 2.50. *Scarf joint above a foundation wall in a traditional solution. Here, the scarf joint is employed to resist tension*

2.2.2.6. Steel-footing connections

It may seem that the simplest way to anchor a post to a concrete foundation, or foundation wall, is to connect them directly, but a direct connection does not protect the timber from moisture, and is therefore especially suitable for greenhouses with a short planned lifetime. The alternatives require adopting some mediation component between the posts and the foundation.

The most satisfactory among the simplest solutions is constituted by anchoring the posts onto timber plates, which in turn are anchored to the foundation walls (see Figure 2.51). The plates can be pressure-treated or made of rot-resistant wood, like larch or chestnut (which requires a greater thickness), and are anchored to the foundation by means of anchor bolts, or strong expansion dowels, and also, they are waterproof with respect to the foundation wall by means of a waterproofing layer separating the two, working as a capillary barrier. However, the strength of the connection between the post and the wooden plate is normally rather small, and therefore the described solution is not primarily used for posts, but for studs, in structures in which the connections are many and distributed, rather than strong and concentrated. To assure a greater resistance to pulling force to the post-to-plate connection, the connection should be aided by some type of steel anchor.

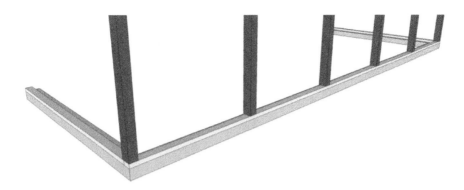

Figure 2.51. *The wooden plate at the foot of the posts is needed to anchor the secondary, light-frame structure. The plate can be anchored to the concrete foundation with screws for concrete, expansion dowels or j-bolts cast into the concrete, most commonly at interaxes of about 1 m or slightly less, depending on the strength of the anchorage and the strength of the plate. On top of this treated lumber plate, a further plate of non-treated lumber may be anchored. In this case, the secondary frame will be anchored to the latter*

To create a stronger connection between the posts and foundation, steel footings are generally used for the posts today because they act as capillary breaks between the foundation wall and the posts, and because they can take a lot of concentrated stress and redistribute it evenly into the timber. The simplest solution of this type is that of a "U" element anchored at the bottom to the foundation, and on the sides, to the column, by means of bolts, nails or screws (see Figure 2.52). However, the moisture break produced by this solution is not great, because the solution does not raise the level of the post foot from the level of the foundation beyond some millimeters. A safer solution is providing the column foot with a connector which, besides resisting extraction, detaches the basis of the column from the surface of the concrete of a distance sufficient to protect the post from the droplets of water bouncing up from the ground during rain or during the watering operations (see Figure 2.53). The connectors, in this case, can have many shapes, and be ready-made or bespoke. The height of the basis of the post that can be considered safe – that is, adequate to protect the bottom of the post from water – is 15 cm. This is at least what is "officially" agreed upon in the technical literature. But, in the absence of frequent and/or copious rain, this height can be much smaller (up to half).

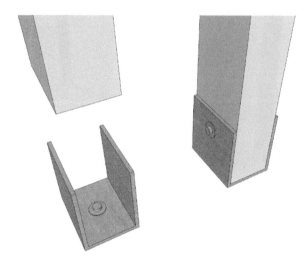

Figure 2.52. *The simplest footing for timber post: a "U" profile bolted to a support. However, in wet environments, this waterproofing break is likely to be insufficient to protect the post from droplets of rainwater bouncing up from the ground (because the timber is too near ground level). This problem is absent if the structural support of a post is located above ground level*

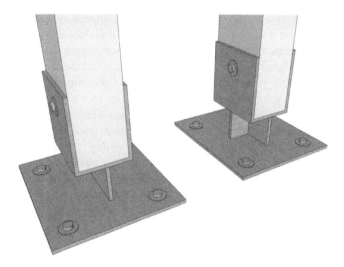

Figure 2.53. *In this case, the steel connector detaches the post foot from the support to protect it from rainwater bouncing up from the ground*

Detaching the post foot from the ground can also be accomplished using standard elements ("C" profile) at a much lower cost. A smart solution to accomplish this (proposed in Alward and Shapiro (1980)) involves the use of two "C"s, that are placed one inside the other, as shown in Figure 2.54.

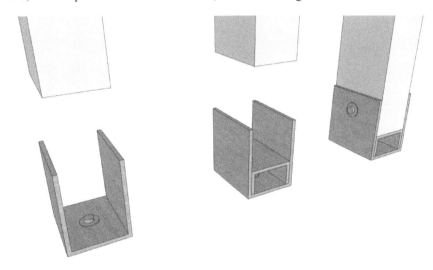

Figure 2.54. *Simple arrangement of U-shaped steel profiles that make it possible to anchor a timber post to a concrete foundation via anchor rods, while keeping the foot of the post safe from water and moisture (Alward and Shapiro 1980). The described connections are best performed with standard stainless steel components, but may also be performed with standard carbon steel, zinced-steel and/or painted steel components*

Another solution to locate the post foot farther from the foundation level, and/or from the ground level, is to sit the post foot directly on the bolts, for stronger capillary action (see Figure 2.55), but this solution is likely to require shear connectors, in order to distribute the stresses within the post's foot, or a substantial number of bolts, or bolts of a large diameter.

A similar result, with stronger fire protection, and different aesthetic effect, can be obtained by means of internal plates (plus, optionally, shear connectors) (see Figure 2.56, left). A solution to increase the load bearable by the foot of the column without requiring steel connectors entails transmitting the loads to horizontal plates (see Figure 2.56, right). In production, this can be obtained in many ways by welding or bolting the horizontal support to the vertical plate. For example, by welding two "T"s onto each other, or by the simple bolting of ordinary pieces like "C"s and "T"s onto each other.

Figure 2.55. In this case, the foot of the posts is resting directly on the bolts

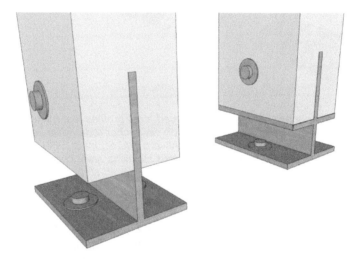

Figure 2.56. In the solution on the left, the timber posts sit on one or more bolts passing through the internal plate, possibly with the aid of shear connectors on the side of the plate. In the solution on the right, the steel footing is characterized by the presence of a horizontal plate, to which the vertical top-plate is welded, and the presence of shear connectors is unneeded. In the market, many types of cup-shaped light-gauge or medium-gauge footings substantially performing a function equivalent to that shown by the footing in the latter example are present. This type of support in greenhouses, however, due to their light weight, is usually justified only when snow loads must be taken into account

2.2.2.7. Connection between columns in highly loaded configurations

If the height of a timber-frame greenhouse allows the columns to be made continuous when they encounter some beam (independently of the fact that the beam is or is not continuous), this opportunity is usually exploited in the design arrangement. But when the greenhouse is very tall (as well as for other possible reasons), or when, for some reason, it is greatly loaded (e.g. in the case of a greenhouse surmounted by a building, a storage space or a pool, or in the case of a full-height post supporting an balcony indoors), there are cases in which the posts have to be divided into two or more trunks, because otherwise they would have to be too long, or at the same time, too long and bulky. In those cases, the posts trunks can be connected with butt joints, or the beams can interrupt the posts. In greenhouses, the vertical loads that are involved in the post-to-beam connections are usually not great, and therefore there is no need to insert an intermediary steel connector between the post trunks, of the kind that may be used, for example, in the heavy-loaded columns of factories built with timber structures. However, the heavy-load solution in question (see Figure 2.57) can be adopted whenever a heavy loading is foreseeable.

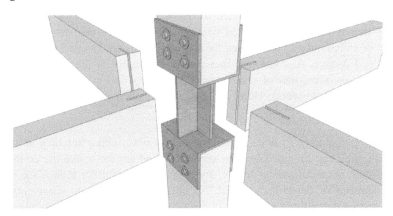

Figure 2.57. *This kind of post-to-post-post-to-beam connection is suitable for heavy loads, but seldom necessary in greenhouses, except in special cases*

2.2.2.8. Choice between nails, screws and bolts for the connections

In many solutions, a choice of screws, nails or bolts can be used. Nails and screws are rather interchangeable and can be used in the same – or very similar – configurations. The diameters used for screws are most commonly 5 or 6 mm, but 4 mm or 7 mm are also used. 4 mm screws are useful for small components, and 7 mm screws are likely to require the timbers to be pre-drilled. The use of screw or nails up to 5 or 6 mm in diameter usually makes it possible to avoid pre-drilling holes for the screws, when ordinary, soft construction lumber is used, but requires pre-drilling with hard lumber. Pre-drilling should be performed with drill bits of a diameter of a couple of millimeters smaller than the screws to be inserted, in order to allow the bit to bite well into the wood.

The diameter of the bolts can be larger than that of the screws or nails in screwed or nailed connections, because a less intimate collaboration between the wood and steel components is required and therefore the steel components can be submitted to a higher concentration of stress. This means that usually, fewer bolts need to be used than screws or nails, and the layout of screws or nails is usually more diffuse than that of bolts. Moreover, the heads and nuts of the bolts can be (should be) accompanied by steel washers, which create a resistance to extraction that is much greater than that of the threads of the screws, which in turn is greater than that of the nails in their lodgment (indeed, the latter only derives from attrition under compression, not shape). Because of these differences, the three solutions have different strengths. The order of resistance is, from the strongest to the weakest, bolted connection, screwed connection and nailed connection.

Because of the greater resistance to tension, the resistance of a structure to swaying that is produced by bolted connections is greater than that produced by screwed or nailed ones, at least when the screws or nails are inserted perpendicularly to the wood surface. A solution to increase the resistance of the screws or nails to extraction is, as seen, that of setting them diagonally and in a mutually divergent arrangement into the wood. However, this solution demands high-level carpentry skills.

A strong advantage of bolts over screws and nails is that when they are used in lap joints, they can be aided by ring connectors, which can make the connections stronger with respect to shear forces, and can therefore further reduce the number and/or diameter of the bolts required. An alternative solution to shear connectors used to reduce the number of bolts is to encase the bolts in hollow steel cylinders, in order to increase the surface contact between the steel and the wood in each cavity hosting the bolts.

The placement of nails, screws or bolts into the timbers must take into account the fact that they should not create an excessive concentration of stress, and therefore should neither be too near to the edges of the elements to be joined nor to each other. As a rule of thumb, in the direction orthogonal to the wood grain, the minimum distance should not be smaller than four diameters from center to center of the components and three diameters from center to edge. Along the grain, the distance should not be smaller than 5 diameters from center to center and 4 diameters from center to edge.

The fact that the distance required along the grain is greater than that required orthogonally to the grain derives from the fact that the presence of connectors along the same grain lines can encourage the wood to split along those lines. A common solution to reduce the risk of splitting the wood along the grain is, as seen, placing the connectors (bolts, screws or nails) so as to make sure that they are not aligned along the grain (see Figure 2.30). But of course, this solution can only be adopted when a connection is not too "crowded" with bolts, screws or nails.

2.2.2.8.1. Operative differences between nails and screws

Nails can be used much more quickly than screws with carpenter's hammers when electricity is not available and electric screwdrivers cannot be recharged and used. Nailing is therefore a solution that is more universal than screwing. Also, nailing can profit from modern devices. Today, however, nails are commonly shot in place with pneumatic pistols, rather than driven with carpenter's hammers. Screwed connections instead have to be executed with screwdrivers to be practically advantageous. As seen, screws are costlier than nails, but screwed connections are stronger than nailed ones, due to the much greater anchoring action they exert in their lodgment, and are also reversible and easy to manage. Due to these advantages, screws have superseded nails in many on-site situations when saving money is not a core purpose.

As seen, the length to be chosen for the nails or screws is most commonly 1 to 2 cm shorter than the overall thickness of the group of elements to be joined. For example, for lap-joining two timbers with a thickness of 6 cm, screws that are 10 or 11 cm long may be used. To join three pieces of timber together, 16 or 17 cm long screws can be used unless the screws or nails are positioned diagonally. The choice of placing the screws or nails diagonally is often coupled with that of placing them in alternated directions orthogonal to each other, so as to distribute the strength on symmetrical sides of the connections and give the connection resistance to tension. However, in those cases, care should be taken to make sure that the screws are not located too near to each other in the wood (see Figure 2.38).

Both nails and screws can be accompanied by thick and large washers, which increase the collaboration with the joined components; however, washers in wood carpentry are not very commonly used in combination with screws or nails. They are with bolts, which bring greater forces, so as to increase the surface of exchange between steel and wood and decrease the stress on the wood. It is useful to couple washers with the screws, in particular in situations in which the liability to extraction of the screws is great – that is, when the tension exerted on the screws is great.

2.2.2.9. *A simplified criterion for determining the number and size of screws, nails or bolts in the connections*

The strength of connections should ultimately be computed analytically. But a preliminary design phase may be based on proportions and simple estimations of the resistance required by the structural components and the connections. The procedure that will now be presented to size the connectors at the connections is simplistic, but useful. The procedure, which can be applied once the dimensions of the main elements of the frame have been defined, can be summarized in the following steps.

1) *Evaluating the amount of timber in the elements to be connected.* After designing the structure, assuming that what is being designed is a hinged, non-bending resistant connection, the dimensions of the elements to be connected has to be taken into account. The most critical dimension is that of the smallest structural element involved in the connection. The evaluations regarding the connection will take that dimension as a reference, because there is no point in making the connection stronger than the weakest of the elements to be connected: a chain is only as strong as the strength of its weaker ring. There is no benefit in making some rings stronger than the others.

2) *Determining the fraction of resistance needed at the connection.* Once the smaller end of a connection has been recognized, a suitable ratio should be determined between the resistance of the smallest components (timbers) to be connected and that of the connection. For example, when this ratio is expected to be 1 (which happens when the connection has to be as strong as the weakest structural component connected through it), to foresee how many lodgments of the nails or screws or bolts are needed, it should be envisaged that the total resistance of the wood put to work into those lodgments should be made as great as that of the weakest structural element of the connection. Therefore, the connectors (again: nails, screws or bolts) should involve approximately the quantity of wood needed to make up that resistance. For example, when the ratio is expected to be 0.7 (70%), the total resistance in the wood called into play in the lodgments of the nails/screws/bolts should be made as great as 70% of the weakest structural element of the connection.

3) *Determining the number of connectors needed to obtain the needed resistance.* The simplified criterion to calculate how much wood is involved in the stress taking place in the lodgment of a connector (screw/nail/bolt) entails taking the diameter of the connector as the frontier of exchange between steel and wood. This can be considered an acceptable simplification for simplified calculations (Cowan 1971). It is the overall resistance of the wood matter (including volume) involved that needs to be measured, and not that of the steel matter (including volume) involved (indeed, the latter is likely to be higher, and no problem is likely to be

derived by it). The number of cylindrical connectors (bolts, screws, or nails) needed can be calculated by dividing the overall thickness of the smallest element to be connected (reduced by the cited ratio) by the size of one diameter. Let us suppose, for example, that, in order to connect a batten 10 cm thick to another piece of timber via external plates accepting a strength reduction of 20% at the connection (which brings the effective depth to be considered to 8 cm), bolts that are 1 cm in diameter should be used, according to this criterion. This means that to obtain the maximum useful (optimal) resistance from the connections (in other words, to make sure that a connection is not weaker than the wood called into play elsewhere in the most stressed wooden component), 1 x 8 = 8 bolts would have to be used. Of course, if the diameter of the bolts were changed, the number of bolts should be changed too.

4) *Defining the placement of the connectors on the basis of their mutual distance and minimum distance from the edges of the timber component.* Once the number and size of connectors have been defined, their position should be defined so as to avoid that the mutual distance of the connectors and their distance from the edges of the timber components is not smaller than the minimum distance allowed (being, as anticipated in section 2.2.2.8, the required distance 4 diameter across the grain from center to center, and 3 from center to edge approximately, as well as 5 along the grain from center to center, and 4 from center to edge, when the connectors are non-aligned along the grain). If the solution resulting from this procedure is not satisfactory, the size and number and/or the position of the connectors should be modified until a satisfactory arrangement is met. As always, the search for satisfactory design solutions is an iterative process involving explorative trials and comparisons between outcomes.

2.2.2.10. *Criteria for positioning the wood grain in wooden elements and their connections*

The grain of a timber component depends on how a tree trunk is cut. With regard to the relations between wood performances and grain direction, it should be considered that the direction along the wood fibers (which are made of cellulose and hemicellulose forming the polymeric structure, and lignin as a binder) is the one in which the mechanical resistance is higher and is also the most dimensionally stable in response to the thermal and moisture variations; the radial direction comes second in terms of both strength and stability (thermal as well as moisture-related), and the direction tangential to the trunk is the weakest and less stable. Other key facts about the resistance of timber are that: (a) the further the fibers are from the center of the tree trunk – trunk core excluded – (i.e. the more peripheral the fibers are), the less strong and dimensionally stable they are; (b) in the axial direction, the resistance to tension of timber is greater than that in compression (almost double); (c) in the direction orthogonal to the fibers, the resistance in compression of timber is much greater than that in tension, and is smaller than the resistance to compression in the axial direction and (d) the resistance to shear of timber along the fibers is the lowest, of the same order of that to tension orthogonal to the fibers, while the resistance to shear across the fibers (perpendicularly to them) is high.

The mechanical resistances in the various directions depend on the species and gradation of timber – that is, on its class of strength, which in turn depends on the growth area of the trunk from which the components derive (see Table 2.3, Volume 4). Of course, those resistances differ for glued-laminated lumber, plywood, OSB and Cross-Lam.

In standard construction lumber, the reason why the resistance in shear parallel to the wood fibers and that in tension orthogonal to the fibers are the smallest is that they depend very much on the resistance of the natural "glue" between fibers.

2.2.2.11. *Criteria for understanding the effect of wood grain orientation in structural components*

The wood grain in connections should be positioned to obtain stability and robustness. The direction of the fibers in a timber component – in modern carpentry, in particular – should usually be decided on the basis of the kind of forces that the component will be called to resist, more than on the basis of the forces in the connection. Then, once the direction of the grain has been defined, the connection, as well as the connection devices used in it, should be designed on the basis of the direction of the forces in the component and the grain direction. From this, in calculations, as a criterion of caution, it is necessary to assume, at least preliminarily, the minimum possible resistance in shear along the grain of the wood hosting the connections. This resistance is called into play when the component is stressed by shear forces parallel to the grain.

The actual resistance of the connected components will depend, in any case, on a much wider set of parameters, such as shape, species, assembling criteria and connection solutions.

2.3. Structural solutions with the primary beams of the frames orthogonal to the front façade

This family of structural solutions is the most rational and simpler for a timber-framed greenhouse, because the repetition of frames (or portal frames) that characterizes it takes place along the width of the greenhouse, so as to require that the main span of the frames lies in the direction opposite to the main axis of the greenhouse, the width. This structural scheme is suitable for both greenhouses attached to walls and self-standing greenhouses.

When a greenhouse is of the lean-to type, its rear columns (or posts, or uprights, depending on the type of structure) can be placed near the wall shared with the building (the timber-frame equivalent of the platform frame in Figure 1.36), or, as an alternative, the shared wall can be called to perform a load-bearing function for the greenhouse (this is the most common solution), in which case, the greenhouse structure can be constituted by half-frames or half-portal-frames in which the uprights on one side are substituted by the wall of the building, or by a self-standing wall (see Figure 2.78). A third option, intermediate between the two cited, is characterized by asymmetric portal frames (see Figure 2.72). The latter option typically occurs when a greenhouse is higher than the building and the building plays the function of support for the uprights connected to it.

2.3.1. *Post-and-beam greenhouses with primary beams perpendicular to the front façade*

This is the most basic and common structure of post-and-beam greenhouses. It can be built with massive lumber or glued-laminated lumber. In this configuration, the posts are usually continuous, and the beams stop at each post. But it can also be that the posts are continuous and double and the beam is single and continuous, or that the post is single and continuous and the beam is continuous and double (see Figure 2.66). The fact that the primary beams are perpendicular to the front façade means that the secondary beams are parallel to it and that the façade frame is anchored to the secondary beam.

As seen, a solution to make the double beams and/or double posts more resistant to buckling can be obtained by joining the double elements at intervals – typically, at 1/3 and 2/3 of the span (see Figures 2.67, 2.68 and 2.69) with timber blocks. When the columns or the beams are double, the lap connections between columns and beams are likely to be aided by shear (ring) connectors.

The type of frame in which the beams do not stop at each post, but rather the posts stop when they encounter the beams, is only common in greenhouses when the only position at which the posts encounter the columns is at roof level (see Figure 2.58). It is, instead, rare that the beams interrupt the posts at an intermediate level. This happens in the case of a multi-story structure.

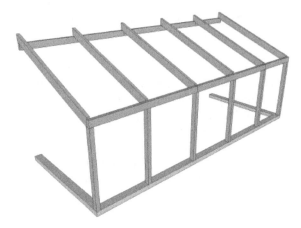

Figure 2.58. *Example of the simplest possible attached timber-frame greenhouse*

NOTES ON FIGURE 2.58.– Each sloped beam rests on the post supporting it, and the posts are joined by a header beam, which is, in turn, connected to the heads of the rafters. In this and the following examples, the connection between posts and the foundation may take place through steel connectors directly anchored to the foundation, or indirectly, onto a plate of treated lumber, in turn, anchored on the foundation wall.

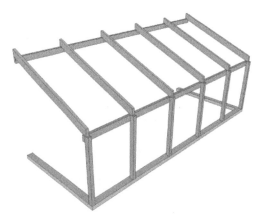

Figure 2.59. *In this variant, the beam at the front is connected to the back side of the posts and can collaborate with them to support the sloping beams of the roof. This solution is not suitable for supporting the sloped beams when they are off-center with respect to the edge beam. The knee between the sloped beams and the posts can be left empty (this is the solution shown in this image) to make room for a gutter*

Timber Frames   113

Figure 2.60. *In this variant, the knees between the posts and the sloped beams are set flush*

Figure 2.61. *In this variant, the post-and-beam parallel frames are connected by a strong plate that is in turn connected to the beams from below. This solution is likely to obstruct less solar radiation in winter than the one shown in Figure 2.60*

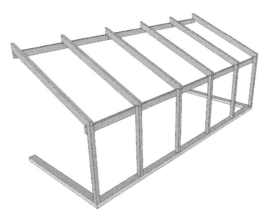

Figure 2.62. *This post-to-sloping-beam connection is strong but more complex than the previous one*

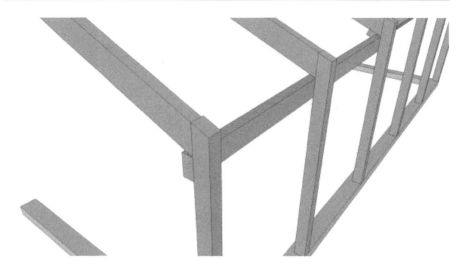

Figure 2.63. *For this version to be advantageous, the post-to-edge-beam connections must be aided by the steel connectors, and the horizontal beam should be suitable to support the sloped beams entirely*

Figure 2.64. *In this variant, the front beam is supported by a recess at the top of the posts, and ends up being away from the vertical plane of the curtain wall. This leaves more freedom (it leaves more options open) to organize the curtain wall*

Timber Frames 115

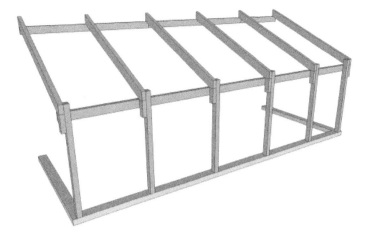

Figure 2.65. *This timber frame (experimental) is entirely built with lap joints, single posts and sloped beams*

NOTES ON FIGURE 2.65.– How the edge beam is anchored to the posts and how the sloped beams are anchored to the posts are both entirely dependent on the connectors that are used (nails, screws or bolts). To reduce the stress at the connections, a corbel for each post is lapped to support the sloping beam. The counterintuitive appearance of this arrangement derives from the fact that the posts and sloping beams do not lie in the same plane. For this reason, both undergo an additional tendency to sway on one side. Therefore, this solution is not suitable for heavily loaded situations.

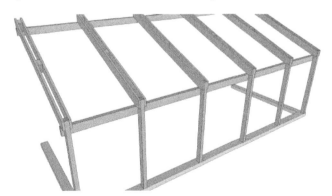

Figure 2.66. *In this solution, all of the connections are lapped, and the sloping beams are doubled. The solution is symmetrical and therefore suitable for heavy loads. The function of the blocks at mid-span of the beams is to reduce the free span of the twin beam components, and consequently reduce the tendency of the twin beam components to buckle under load*

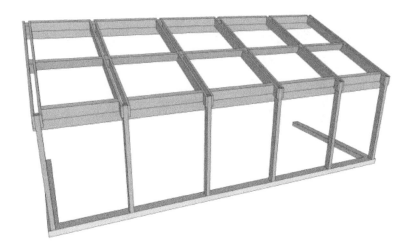

Figure 2.67. *In this variant, wooden stops have been located above the edge beam, the top beam and at mid-span of the sloped beams, in order to improve the constructability of the roof – specifically, to distribute the forces to both the sloped beam components*

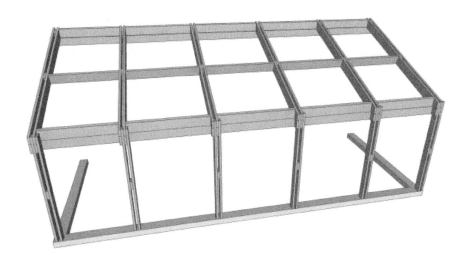

Figure 2.68. *In this lap-joined variant (experimental), both the columns and the beams are doubled. This allows us to use posts of narrower section, but this also generates a tendency to buckle under load in both the posts and the sloped beams, due to the asymmetry of the post-sloped beam configuration*

Timber Frames 117

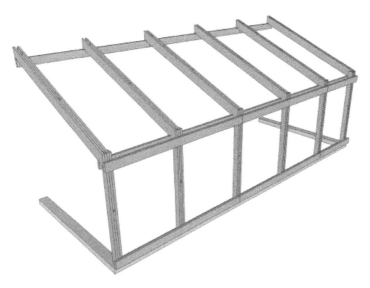

Figure 2.69. *In this fully lapped variant (experimental), the header beam has also been doubled. To facilitate the assembly, subsequently, the façade using studs/ mullions, an edge beam has been added at the bottom of the front façade, on the same plane as the outer header beam component*

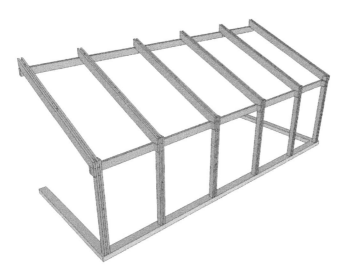

Figure 2.70. *In this fully lapped double-post and double-sloped-beam variant (experimental), a side-corbel at the head of each post has been added for each double beam. The frontal beam at the back of each post aims to connect the frames transversally, without supporting any load from the roof*

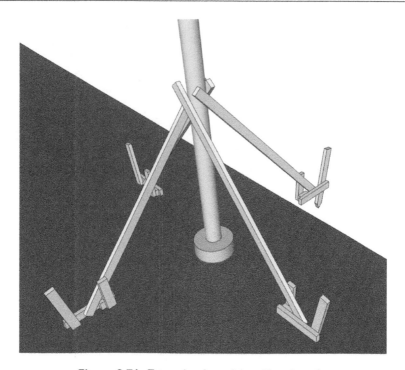

Figure 2.71. *Example of provisional bracing of a timber pole standing on a concrete foundation pad*

NOTES ON FIGURE 2.71.– This bracing is usually obtained with three props, but can also be obtained with four. This propping is necessary when a post-and-beam structure is built by setting up the posts, and in a second operation putting in place the beams on them, rather than assembling the frames composed of posts and beams in advance, at the ground, then tilting them up. In the latter case, the pre-assembled frames, rather than the single posts, are what have to be provisionally propped.

2.3.2. *Trussed post-and-beam greenhouses with primary beams perpendicular to the front façade*

This version of the post-and-beam structure in question is, until now, seldom used, but it may be used more in the future because it has great potential. Indeed, it constitutes a rational solution to extend the dimensional possibilities of sawn lumber beyond those of single wooden elements, and trussed frames are easier to build and connect than massive portal frames.

The solution is based on two assumptions: (a) that both the post and the beams are framed, and are therefore of greater resistance than the single timbers composing them, and (b) that the framed posts and beams can be connected by placing them side by side (and not one of top of the other), in such a way that the main plane of one does not coincide with the main plane of the other. This is done on the assumption that the secondary bending moments and the torque developing due to the asymmetry of this lapped truss configuration are small compared with the principal bending moments and shear forces (see Figure 2.79). That said, by doubling the columns or the beams, it is possible to make this frame configuration symmetric (see Figure 2.80), so as to reduce the parasitic torque resulting from lapping the beams to the posts.

The described solution can be obtained either using wooden elements of a section suitable for light frames (this is the most common situation) (see Figures 2.79 and 2.80), or using timber sections of a dimension suitable for timber frames. The main thing to consider here is that, once the frames of the post (in portals, also called "uprights") and beams are assembled, the trussed frame will work as a timber frame would, in the overall structural scheme of the greenhouse.

The connections between posts (uprights) and beams in the described scheme are mainly lapped and distributed at each position at which the trussed uprights and beams encounter each other side by side. These lapped connections may be executed by bolting, with or without shear connectors, or by screwing or nailing – provided that the stresses allow for that and the configuration of the structural element makes a satisfactory (i.e. even and well-spaced) distribution of the screws/nails possible.

In the places where the trussed uprights and trussed beams encounter each other at a certain distance (i.e. commonly, the thickness of one, or sometimes two, pieces of timber), the gap between them may be filled by intermediate blocks. The gaps are likely to be present, especially when the trussed beams and uprights are themselves constructed with lap joints. The necessity of filling the gaps by means of intermediate blocks is not likely to be present when the sides of the trussed elements (chords, diagonals and uprights) are flush with respect to each other. This is likely to take place when they are designed with continuous chords and joined with bolts or screws/nails oriented along the plane of the truss, or with tooth plates.

A further consequence of the fact that the punctual connections between trussed uprights and trussed beams are distributed and spread across a wide lever distance is that those connections between uprights and beams can be easily made rigid, bending-resistant. This makes those structures self-braced along the plane of the trusses, and also qualifies them as portal frames (i.e. frames having rigid, bending-moment-resistant connections between uprights and beams), even if they are not built like ordinary trussed ones – considering that in the latter there is usually continuity between the trusses of the uprights and the trusses of the beams.

2.3.3. *Portal frames perpendicular to the front façade*

Portal frames, as said, differ from frames because of the continuity, and the derived rigidity, between posts and beams. When that connection is performed off-site and in a factory/carpentry, for portal frames of a small size, both the structural scheme of the double-hinged portal frame (characterized by hinges at the feet of the posts) and that of the triple-hinged portal frame (characterized by an additional hinge in ridge position) can be used, while for portal frames of a large size, there is almost no alternative to the triple-hinged portal frame, because pre-assembled double-hinged units would be too large to be transported. The two halves of three-hinge portal frames are more maneuverable and can be transported separately, to be assembled on-site.

The main technical difficulty in constructing the portal frames is assuring bending-moment resistant connections between uprights and beams, and this is particularly evident when those connections have to be performed on-site. The accomplishment of this task is most commonly pursued through one of the strategies presented in the following sections.

2.3.3.1. *Timber portal frames braced with timber diagonals*

A common strategy for constructing portal frames is to make the resistance to bending between uprights and beams derive from diagonals joining the uprights to the beams. Those bracings may be suitable for working both in tension and compression, as it happens in lap joints when the connectors (bolts, screws or nails) work in shear (see Figure 2.72), or they may be suitable for working in compression only, as it happens, for example, with bracings connected with traditional tenon-and-mortice joints (see Figure 2.26), or they can work in tension only, when the diagonals are constituted by steel cables or rods. The strongest of the three solutions is when the diagonals can work both in tension and compression, because both diagonals are always called into play whatever the verse of the horizontal force along the plane of the frame, while only one diagonal at a time is called into play when the diagonals can work only in compression or tension (see Figure 2.104).

Timber Frames 121

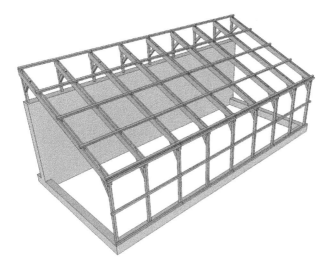

Figure 2.72. *When the uprights of the frames are braced against their beam, a portal frame is obtained. When one of the columns of each portal frame is completed by the wall, the portal frames are asymmetrical. Portal frames are usually connected by purlins onto the roof and onto the wall, more often than by concentrated beams. This simplifies anchoring the light-frame parts constituting the curtain walls and the curtain roof*

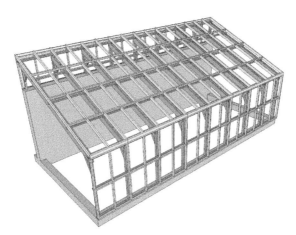

Figure 2.73. *This case is a follow-up of the previous one. The timber-frame elements are shown in orange, and the light-frames ones in beige. In this scheme, the purlins belong to the timber-frame structure, and the rafters/studs/mullions constitute the primary light-frame elements. The transoms and the studs/mullions constitute the secondary light-frame elements*

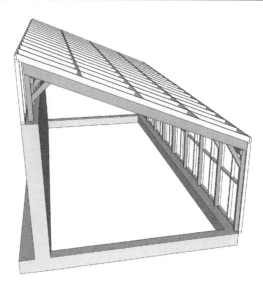

Figure 2.74. *The light frame enclosing a timber-frame portal can be constituted by purlins as primary elements and mullions as secondary ones. In this case, the purlins belong both to the timber-frame system and to the light-frame one, and also work as transoms*

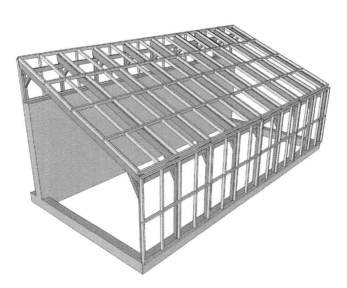

Figure 2.75. *In this case, the portal frames are connected through concentrated beams (edge-beams), and not by the purlins. The purlins here only work as transoms: they are the primary elements of the light-frame system, but they do not "belong" to the timber-frame one*

2.3.3.2. *Trussed, light-framed or timber-framed portal frames*

Another strategy is building the portal frames by making the lap-joined elements (uprights and beams) constituting the portal frames "flow" into each other, establishing structural links between each other spaced widely enough to create a lever arm sufficient to make the connection between uprights and beam resist the bending moments at play. This, as seen, can be done by using wooden elements of a section suitable for light frames (which is the most usual situation) (see Figures 2.76 to 2.78), or using timber elements of a section suitable for timber frames (see Figure 2.72). In any case, at the constructional level, most often the portal configuration entails putting great care into constructing the first frame, and producing the next frames more quickly, by copying the first one.

One solution that makes it possible for the uprights and the beams to "flow" into each other is doubling one of the two, and connecting the two with lap joints, like in the following figures (from Figure 2.76 to Figures 2.78, 2.80 and 2.81) (as well as in the solution experimented in the case-study greenhouse shown in Volume 4, section 3.2). An example of this solution is the greenhouse shown in Volume 4, section 3.2. In that case, the greenhouse structure is constituted by semi-portal frames sitting on the back of the concrete wall, in which the columns and the beams are made continuous by doubling the columns, keeping the beams single, and mutually connecting them by means of screwed lap joints.

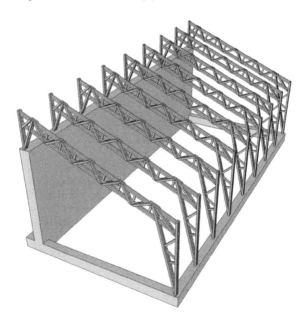

Figure 2.76. *In this case, the portal frames are trussed*

NOTES ON FIGURE 2.76.– This structural scheme resembles that of the solar greenhouse shown in Volume 4, section 3.2, with the differences that (a) here, the diagonals and the chords of the trusses are constructed using only one type of wood section, while in the greenhouse, the chords are made with wooden battens and the diagonals are made with wooden plates; (b) here, the diagonals are double and lap-joined to the chords and the upright are single and butt-joined to them, while in the greenhouse in Volume 4, section 3.2, both the diagonals and the uprights are double and lap-joined to the chords

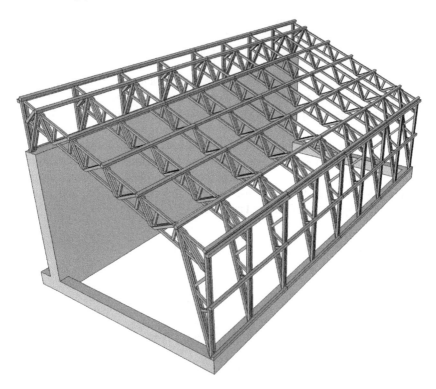

Figure 2.77. *In the case of a trussed-portal frame greenhouse, if the chords of the truss beams are not sturdy, the purlins should better transmit their loads at the nodes of the trusses – that is, at points where the diagonals meet. This avoids stressing the upper chords in bending between the connections of the diagonal by the local loads*

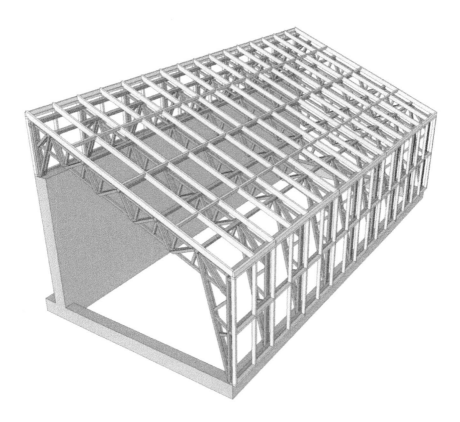

Figure 2.78. *A conventional light-frame roof and façade can be laid out on the purlins, which in turn connect the portal frames*

Another solution is that the uprights and beams of the trusses are constituted by ready-made separate elements and that the connection between each upright and each beam is lapped, so as to obtain a symmetric compound when double uprights (see Figures 2.80 and 2.81) or double beams are used, and to obtain an asymmetric compound when single uprights and single beams are used (see Figure 2.79). In the latter case, the uprights and their beam lie on different planes, and secondary lateral bending moments and torque appear in the portal frames, to the point of making them sway under load (usually, almost imperceptibly – which makes the solution feasible).

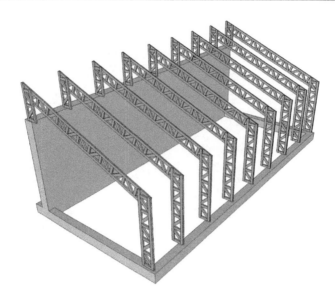

Figure 2.79. *These trussed portal frames (experimental) are built by asymmetrically lap-joined truss beams and columns (each of which is of a type in which chords, uprights and diagonals are constructed so as to have the chords, uprights and diagonals flush with respect to each other). The condition to assure the strength of these portal frames is that the connections between columns and beams are sufficiently diffuse, and therefore strong. These connections should not make the connectors (bolts, screws) too densely packed near each other in the zones of overlap between columns and beams*

Figure 2.80. *View of a trussed portal frame similar to that used in the previous example, but symmetric along its plane (experimental). This is because here, the uprights are doubled (and the beam is single)*

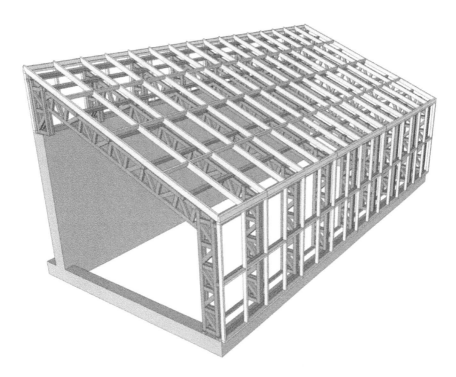

Figure 2.81. *The light-frame enclosure arrangement (experimental) may remain substantially unchanged regardless of the configuration of the underlying portal frames, provided that the external contour of the portal frames does not change*

2.3.3.3. Timber portal frames with the uprights rigidly connected to the beams

Another possible timber-frame portal construction solution is constituted by making the uprights and their beam "flow" into each other in a more integral way, diffusely, alternating their layers in an interlocking fashion. The result may be similar to that in which the uprights and beams are massive, non-trussed and the only triangle in play is that of the two bracing diagonals at the connections between columns and beams, like in Figure 2.75, or the result may not have bracing diagonals at all.

The latter solution is most commonly obtained in one of the following ways.

2.3.3.3.1. Massive portal frames with the post-beam connections formed by bolts arranged in torque-resisting shapes

One technical option is building the uprights and beams in layers, making the connection between them wide enough to make it bending-resistant thanks to the overlap between rigidly connected layers. A connection solution allowing this is constituted by bolts, possibly aided by shear ring-connectors, or screws or nails, set as far away as possible from the center of rotation (see Figure 2.82 and Figure 1.18 of Volume 4). The "layers" composing the uprights and beams, in that case, can be made of large solid sawn planks, or of thick (in the order of 2.4 cm) plywood or OSB panels in multiple alternated layers.

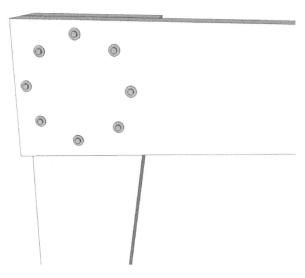

Figure 2.82. *Example of bending-resistant bolted connection between an upright and a beam in a timber portal frame*

2.3.3.3.2. Boxed portal frames enclosed between plywood or OSB panels

Another option is enclosing a discontinuous frame positioned at the edge of the portal shape with large plywood or OSB panels cut to shape and joined in an overlapped fashion (see Figure 2.83). In that context, the role of the inner "skeleton" is mainly to assure resistance to swaying in bending.

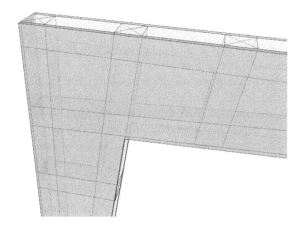

Figure 2.83. *Example of "boxed" portal frames built with plywood panels outside and sawn timbers inside (here visible in transparency). The external panels may be laid out in multiple layers*

2.3.3.3.3. Glued-laminated portal frames

A further option is using glued-laminated portal frames. In that case, the rigidity of the connection between uprights and beams can derive from how the planks are glued with one another within the elements. The combination possibilities to obtain this are huge. These types of structural elements can be chosen from a catalogue or bespoke-made in a factory, on the basis of the indications of the designer – but commonly are not designed by the greenhouse designer.

2.3.4. *Spans of the secondary structural elements in greenhouses having the principal beams orthogonal to the main façades*

There are two solutions for framing the secondary spans of timber frames: (a) using secondary beams connected at, or near, the zones where the uprights meet the primary beams (either by interrupting the beams or by keeping them continuous) or (b) diffusely connecting the frames, with distributed secondary beams (see Figures 2.72 and 2.77), in a purlin-like fashion. The latter solution is more often used with portal frames, while the former is more often used with frames. In the case of distributed beams (case b), they are very often directly constituted by the purlins. In the case of concentrated secondary beams (case a), the elements bearing the panels can be orthogonal to the beams or parallel, depending on the orientation of the transparent panels. In other words, the secondary frames depend both on the orientation of the structural frames of the greenhouse and on that of the transparent panels.

2.3.5. *Frames or portal frames, solid or trussed, parallel to the front façade*

In the case of the timber-framed solution characterized by frames parallel to the front façade, the width of the greenhouse is a multiple of the span of the frames, and it is inevitable that some posts will be located in the middle of the greenhouse. It may be added that in this structural scheme, the primary beams cast a substantial shadow into the greenhouse. These facts reduce the attractiveness of this solution for solar greenhouses, especially when the primary beams are truss, non-solid beams. Truss beams are indeed more slender than ordinary ones, and produce more inter-reflections between their members, when they are of clear colors. This limits the penalization in terms of lighting levels.

The choices available for the construction of the frames, in this case, are very similar to those that have been seen for the frames perpendicular to the front façade, except for the fact that the beams of the frames have no slope and the roof slope is usually obtained by using frames of decreasing height from the back to the front of the greenhouse (see Figure 2.84). The solution in question is especially adequate for the construction of solar greenhouses that are both deep and wide, but for greenhouses that are wide, but not deep, the utilization of frames (or portal frames) orthogonal to the front façade is usually more advantageous.

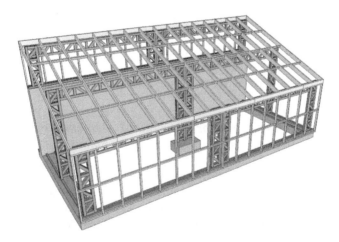

Figure 2.84. *Example of trussed portal-frame structure with portal frames parallel to the front façade (experimental). (Note that this figure is not complete of the bracings in the direction orthogonal to the portal frames.) This is apparently not a common structural scheme in solar greenhouses, neither when the primary structural elements are trussed, as shown in this figure, nor when they are not trussed. However, many types of extensive greenhouses are based on a similar scheme*

The described structural scheme is not common in solar greenhouses, but is common in extensive commercial greenhouses with multiple gables running across the front façade. The ridge-and-furrow Venlo type of commercial greenhouse – constructed in steel – can be classified as belonging to this type (see Figure 2.85). Ridge-and-furrow commercial greenhouses are very common at all latitudes, and not only in the Venlo variants.

The principal difference between multi-bay greenhouses in temperate climates and in warmer climates is that in warmer climates, greenhouses tend to be taller, because more height is advantageous to improve wind ventilation and stack ventilation and increase the stratification of air temperatures.

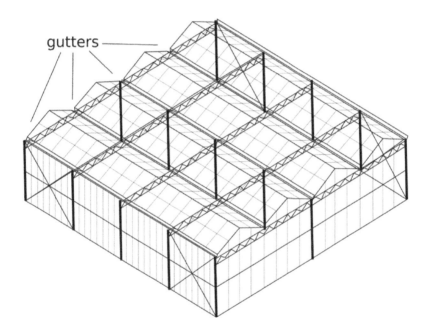

Figure 2.85. *Scheme of a Venlo-type greenhouse, characterized by ridge and furrows on the roof and frames parallel to the front façades*

The utilization of a single order of portal frames parallel to the main façade is also possible for constructing deep and narrow greenhouses. In those cases, three-hinge portal frames may be used when a gable is needed in the front façade. Three-hinge portal frames may also be utilized in the case of multi-bay greenhouses with a series of saw-tooth-like ridges and valleys running perpendicularly to the main façade.

Another structural solution in which the principal beams are most commonly parallel to the front façade is the so-called pole construction (described in the next section), in which the columns are built as poles rigidly cantilevering from the ground.

2.4. Pole construction

Today, the timber pole construction technique has a small "market share" due to drawbacks deriving from its core feature: each pole (column, post) constitutes its own foundation.

The most common pole construction type is that of many North American barns, but the technique can also be used for greenhouses. The main difference between the framing of a barn and that of a greenhouse is that some parts of the barn may be characterized by a floor above the ground level (and therefore by a system of beams supporting a framed floor above the ground, besides the roof, plus possibly intermediate floors), while a greenhouse usually lacks a framed floor above the ground (see Figure 2.87).

The technical problem of the pole construction technique is constituted by the difficulty of assuring durability to the poles, because they have to be embedded into the ground, to "cantilever" rigidly out of it. Indeed, the fact that the poles are embedded into the ground exposes them to moisture, bacteria and insects from the soil, and this shortens their lifetime substantially.

On references

The book by Kern and Kern (1981) is an inspirational reference based on a top-quality realized project. The book by Wolfe (1980) is a solid handbook about the same topic.

2.4.1. *Treating timber poles for a longer life span*

Several solutions to increase the durability of ground-embedded timber poles exist, but none of them are free from issues. The options are as follows.

2.4.1.1. *Using moisture-resistant wood species*

The most straightforward solution is using timber milled from especially durable tree species – typically, larch (a slow-growth, high-altitude tree) or chestnut. However, the technique of pole construction requires a kind of usage of timber that is far less parsimonious than that used ordinarily in timber construction. While the use of rot-resistant timber is indeed only ordinarily aimed at the protection of wood in contact with the ground, or with concrete walls that are in turn in contact with the ground (typically, the plates surmounting foundation walls), in pole construction, instead, the whole length of the pole has to be of a rot-resistant species. It must be added that the life span assured by rot-resistant timber is not infinite. These timbers just last substantially longer. From this, it descends that, despite its advantages, larch should not be considered as a large-scale generalizable solution for making timber durable in conditions of contact with the ground; and pole-frame construction is not the most suitable kind of construction technique for leveraging on the features of larch.

2.4.1.2. *Charring the poles' ends*

The most classical protection solution for timber poles is to make them rot-resistant by charring one of their ends with fire. This is also the traditional method already suggested by Vitruvius (1998, originally published 30–20 BCE) in his book *De Arquitectura*. The main limits of this solution are that the level of protection obtained from it is uneven, depending on the tree species involved and the skill level of the workmanship, and that the durability of the poles is difficult to predict, due to uncertainty.

2.4.1.3. *Painting the poles' ends with bitumen or other waterproofing admixtures or preservatives*

Another solution is painting the ends of the poles with some waterproofing and preservation composite admixture. There are two main problems with this solution. (a) Painting the poles does not usually assure a sufficient penetration of the preservatives into the wood, and therefore does not produce a sufficiently strong and durable preserving action. (b) To be suitable for preservation, the composite admixture to be used as a varnish must contain some antibacterial and antifungal principles, and is therefore likely to be, to some degree, poisonous for humans. This should be put into the context of the fact that, being the compound admixture painted onto the ends of the poles, it can be diffused into the ground, with dangerous effects on human health and the environment.

The preservative used to preserve the timbers is naphthenate.

2.4.1.4. *Pressure-treating the poles*

The usual solution to make sure that the poles are durable is using pressure-treated lumber: lumber heated in an autoclave under pressure after being soaked or painted with preservatives suitable for contrasting the growth of bacteria and fungi, and absorbing those substances in-depth due to the pressure. Pressure-treated timber is usually recognizable because it appears greenish and unattractive. But their greatest drawback as foundation poles is, like the simply painted ones, that when they are embedded in the soil, they release the substances they have been treated with. This should be avoided, especially when the soil is used for growing edible plants.

2.4.1.5. *Using roasted lumber*

The most promising and, in perspective, advantageous solution is "roasting" the lumber in a kiln for a long period, of hours, so as to bring it to a certain degree of mineralization, which increases its strength and durability. The technique of roasting lumber (torrification) is being applied more and more in the construction industry with very promising results. Roasted lumber is included in the broader category of thermally modified lumber.

A very attractive aspect of this technique is that it makes it possible to convert normal timbers into a very durable material without the addition of extraneous substances and without risks of poisoning the soil. The current drawback of the technique is that it makes the timber embody a great deal of additional energy, because the heat treatment requires an oven to be brought to high temperatures (around 500°C) for a long time (of the order of 24 hours). The problem is aggravated by the fact that in the present state of things, the energy that is necessary for the roasting process is currently still being produced by combustion, which increases the level of carbon dioxide in the atmosphere, and therefore the greenhouse effect at a global level.

On references

In the article by Bourgois and Guyonnet (1988), the features of torrified wood are analyzed scientifically.

2.4.2. *Solutions for cantilevering the poles from the ground*

2.4.2.1. *Driving the poles*

A solution for "cantilevering" the poles from the ground is driving them into the ground with mallets. However, mallets are heavy machinery requiring skill and experience. An interesting example of application of malleted poles is that of the Loblolly House, by Timberlake and Kieran (2008)[4]. An added advantage of this solution derives from the fact that the malleted poles exert a compacting action into the ground, increasing its load-bearing capacity. The load-bearing capacity of

---

4 Besides being presented in the book cited in the reference, the project is presented on the website of the design firm: https://kierantimberlake.com/page/loblolly-house.

malleted poles can derive from the attrition with the soil, from the fact that the foot of each pole stands on load-bearing soil, or both.

On references

The project of the Loblolly house is presented in detail by its authors Timberlake and Kieran (2008) in a dedicated book.

2.4.2.2. Embedding the poles in ground pockets

The solution of embedding the poles in ground pockets (holes to be filled afterwards) is the most practical one and is executable with light equipment without requiring great skill and expertise. In this case, the load-bearing capacity of the poles has to derive from the fact that the foot of each pole stands on load-bearing soil, rather than from attrition with the soil. The first step in executing this solution is creating a lodgment – a hole – for the pole in the ground (see Figure 2.86). The hole in the ground should have vertical walls and arrive at a depth at which the ground is compact and load-bearing. The hole diameter should be about 10 cm larger than that of the pole, and it should not be greater than its depth. Preparing the hole entails the following steps.

1) Creating the hole

Nothing prevents creating the hole by digging the soil with a hoe, but the operation is tiresome and time-consuming. More efficiently, the holes can be created using augers, manually driven or motor-driven.

2) Creating the base

The second step is creating a base for the foot of each pole. This is needed to prevent the foot of the pole from sinking into the ground once the pole is loaded. The simplest (and weakest) base for the pole is constituted by compacted gravel. A stronger support layer is constituted by a thick concrete screed, better if placed above a gravel bed for drainage. Ideally, the thickness of the screed should be at least sufficient to assure that the angle of inverse cantilever of the screed from below the pole base is at least 45° (which entails that the protection of the screed, out of the pole, should not exceed the thickness of the screed itself). An alternative to the concrete screed is a flat rock of suitable dimension, but this solution makes it difficult to control the height of the support with precision, if the flat rock is raw.

3) Positioning the pole

The next step is to position the pole at the center of the hole and brace it in the vertical position with provisional (wooden or steel) stakes (usually three, but possibly four) driven into the ground (see Figure 2.71).

4) Compacting the filler

The final step is filling the void parts of the hole with medium-sized gravel – ideally, at least 3 cm in diameter, and preferably round-shaped, for proper water drainage – and tamping it heavily for compaction, taking great care not to move the position of the pole in the operation. As an alternative, the pit could be filled with lean (low-cement-content) concrete, but in this way, the timber post becomes less durable, due to moisture and drainage, which is worse.

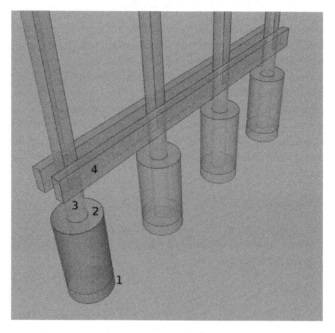

Figure 2.86. *Basic pole construction scheme. Legend: 1. rock, crushed rock or concrete pad layer; 2. gravel; 3. pole; 4. timber girder*

2.4.3. Solutions for connecting girders or beams to the poles

The solution of locating the girders or beams in between the poles in such a way that the beams interrupt themselves each time that they encounter a beam is rarely adopted in pole construction, because this would require having butt joints between all the poles and all the girders/beams, which would make the construction more complicated. This is also because the spacing of the poles in pole construction is usually not wide, and the interrupted beam trunks in that scheme would be rather short. The fact that cantilevering the beams would be more difficult can be added to this.

The most common solution is, rather, that of fitting the girders/beams onto pockets carved into the poles (see Figure 2.88) or onto corbels connected to the poles, or asymmetrically, on one side, or (to attain greater strength and durability), usually, symmetrically, on both sides (see Figure 2.97). In all cases, the connections are usually performed by bolting and can entail the use of shear connectors, especially when the joints are not of the "sit" type.

2.4.4. *Pole greenhouse construction*

In the case of greenhouses, the pole construction is conceived around the poles, to which, basically, everything else is hung. Such a kind of construction is usually characterized by girders (double or single) located above the ground level, laid out in one direction, and lapped to the beams, so as to support everything else. But as an alternative, the construction may also be organized around distributed horizontal purlins, and may lack a ring beam (edge girder) above the ground level. The girders are most commonly double and sit on the pole, but they may even be asymmetrical and sit on the outside face of the poles (for ease of construction of the façade) or (rarely) on the inside face. The only mandatory continuous beam in this arrangement is the top one, which closes the frame at the side opposite to that of the ground (see Figure 2.95).

The secondary structure combined with the described arrangement can be based on horizontal purlins anchored to the poles (see Figure 2.101) or on vertical studs/mullions anchored to ring beams at the bottom level and top level of the greenhouse (see Figure 2.102). In the former case, when the transparent panels adopted are of a small size, they can be fixed on studs/mullions (in turn, anchored to the purlins and combined with transoms) or on the purlins (in turn, combined with segments of mullions between them). In the latter case, the transparent panels are commonly fixed directly on the studs/mullions and combined with transoms.

It is interesting to note that the principles that have been described with regard to pole greenhouse construction can also be applied successfully to many other kinds of non-timber-frame and non-light-frame solutions, like masonry frame panel structures, reinforced-concrete frames, vertical slab structures, hot-rolled steel structures, non-light-framed cold-rolled steel structures, or even machine-cut OSB-panel frames. One example of such a possibility is the Wikihouse (see: https://www.wikihouse.cc), built with plywood panels (Priavolou and Niaros, 2019).[5]

---

5 The interesting rationale of the Wikihouse and its implementation are presented on the website: https://www.wikihouse.cc.

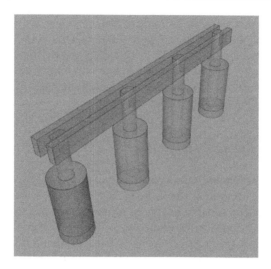

Figure 2.87. *An interesting thing about the pole foundation is that it does not even need a trench to be dug. It is therefore very "light" on the ground*

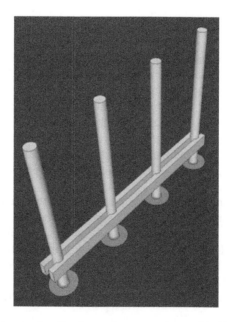

Figure 2.88. *The pole construction is most commonly used in North-American barns, but it has the disadvantage that the posts are all-height long, and their feet have to be rot-resistant (e.g. by pressure-treating)*

Timber Frames   139

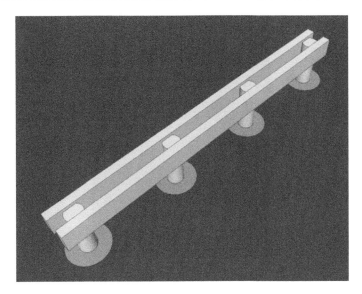

Figure 2.89. *A modern way to use the benefits of pole foundations escaping the related drawbacks is that of making the rot-resistant poles only long enough to reach the girders, and re-start a new kind of construction – like, for example, a light frame construction – from there upwards. In a greenhouse, this could be done when the greenhouse, for some reason, needs a floor detached from the ground, or when an additional ground-embedded frame is provided to separate the soil outside the greenhouse from the soil in the greenhouse*

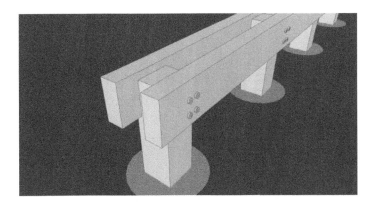

Figure 2.90. *The ground-embedded poles can be worked out to support a double girder (main beam) in notches at their sides, or on corbels. As an alternative, the double girder may sit entirely on bolts. In that case, wood-to-wood shear connectors are very beneficial*

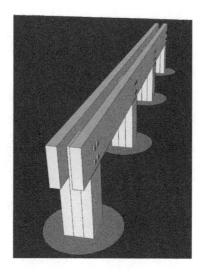

Figure 2.91. *The ground-embedded poles can also be composed of smaller planks, placed side-by-side; however, the risk exists that, in this way, the timbers, being thinner, will be less self-protecting, and therefore less durable. This should be taken into account in the choice of the type of timbers used in the foundation*

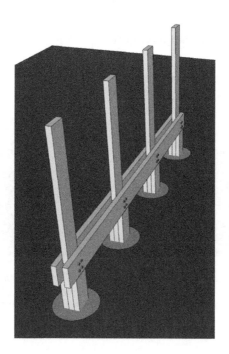

Timber Frames 141

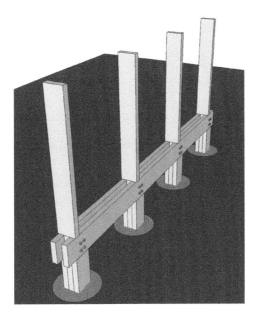

Figure 2.92. *The poles in a pole foundation do not necessarily have to be full-height. They may even be just tall enough to support the floor platform (present), or the double girder. This makes it possible to use pressure-treated lumber only for short timber lengths*

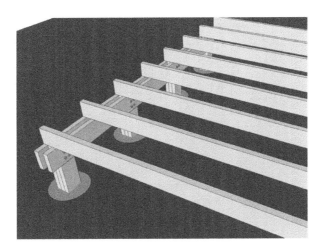

Figure 2.93. *In this example, floor joists are laid out both onto the foundation posts and on the edge double girder*

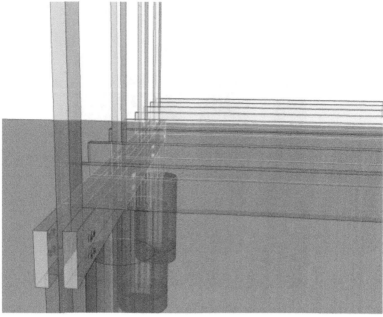

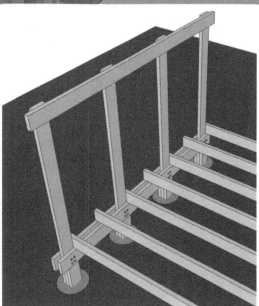

Figure 2.94. *In this example, the greenhouse frame is parallel to the façade. This makes it possible to create a certain continuity between the foundation poles and the greenhouse posts, even when they are not executed in one piece*

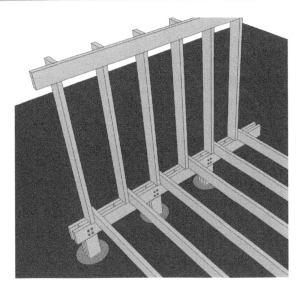

Figure 2.95. *In this case, the posts are conformant to an ordinary light-frame configuration, so as to obstruct as little solar radiation as possible*

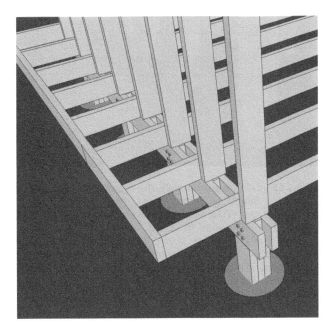

Figure 2.96. *This light-frame floor cantilevers out from the girder*

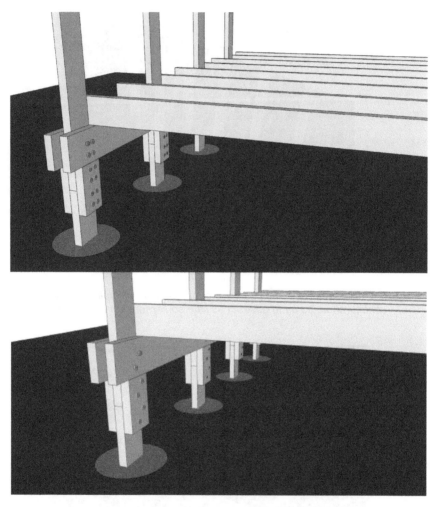

Figure 2.97. *In this example (experimental), the foundation posts are not composite, but constituted by one single embedded post, assisted by a pair of corbels for supporting the double girder. Two alternative bolting schemes are shown in the two images. This solution makes sense for a light assembly. Its advantage is that the footing can be substituted one by one, if necessary. However, care should be taken with regard to the durability of this configuration when ordinary timbers, rather than pressure-treated timbers, are used in contact with the concrete, due to the small size (and lack of redundancy thereof) of the involved timbers. Their small size increases the risk of failure due to rotting in moist conditions*

Figure 2.98. *In this variant of the previous solution, the principal beams, onto which the joists are laid, are orthogonal to the girder. Note that the beams are placed asymmetrically with respect to the posts. This is possible due to the small weight involved, which makes greenhouse structures more "forgiving" of asymmetries than those of ordinary buildings*

Figure 2.99. *In this case (experimental), the double beams are symmetric with respect to the posts. Like in the previous case, three orders of horizontal elements are used in the floor, for greater spans: girder, double beams and joists*

Figure 2.100. *In this variant, because the joists do not sit on top of the double girder, but are connected to it at its same height level, it is necessary to connect the two components of the header double girder via intermediate blocks at each rafter, in order to avoid only half of the girder being called to resist the load*

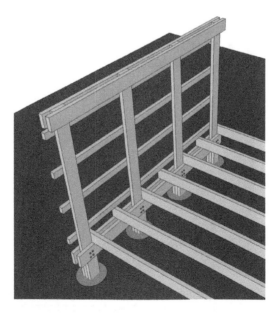

Figure 2.101. *Example of arrangement combining the pole construction with purlins to complete the envelope*

Timber Frames 147

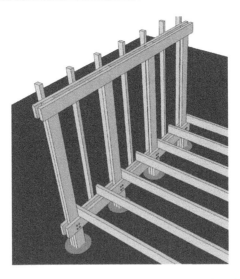

Figure 2.102. *Example of arrangement combining the pole construction with studs/mullions to complete the envelope*

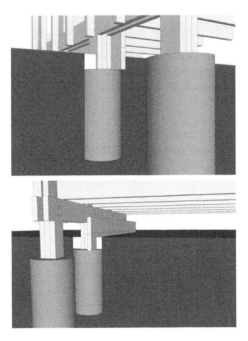

Figure 2.103. *Arrangement combining a pole configuration with a wooden "skirt" of treated lumber to separate the soil of the greenhouse from the external soil. In white above the foundation pile: gravel*

2.4.4.1. *Platform-on-poles foundation construction*

An interesting solution for using poles while limiting their length (so as to decouple the size – and, possibly, the interaxis – of the poles constituting the foundation from the size – and, possibly, the interaxis – of the structural elements above the foundation level) is that of making the poles short, so as to make them support the girders (see Figure 2.89) and the greenhouse structure above, which may or may not include a floor. Those girders can support the structure above them, whether it is a light frame (see Figure 2.95) or a timber frame (see Figure 2.94). The solution in question has been thoroughly presented in Anderson (1969). In the case of the technique described in that publication, the pole trunks bear a light-frame wooden structure onto which a platform-frame structure is built. Now, that solution is not suitable for most greenhouses, because most greenhouses are not raised above the ground. This means that continuous foundations are needed at their perimeter. The function of the foundation walls is indeed in the case of greenhouses, not only that of supporting the structure above, but also dividing the ground enclosed by the greenhouse from the ground outside it, as well as supporting the rigid insulation layers that decouple those two separated ground masses thermally. These functions have to be performed with additional components, in the described technique (parts of masonry walls, frames, panels) (Anderson 1969–2005).

2.4.4.1.1. Enclosing the ground in pole-and-trench foundations

In this section, an experimental dry foundation solution for continuously enclosing the ground despite adopting a pole foundation and without requiring the construction of a foundation wall is presented. The technique entails the following procedural steps:

1) *Digging the trench.* A continuous trench should be dug at a depth suitable for bearing a concrete foundation wall of equivalent functionality.

2) *Setting the poles.* Holes for foundation poles at the bottom of the trench should be drilled, and a pole should be inserted into each of them, at a height suitable for supporting a beam above the ground, following the procedure for creating pole foundations (as mentioned above).

3) *Placing the insulation and the impermeabilization layer.* The outer edge of the trench should be completed with insulation panels and with the waterproof layer externally. To improve the insulation, the trench may be made larger and then filled in two phases to make it possible to include a horizontal insulation layer below ground, outside the vertical insulated panels.

4) *Filling the trench.* The trench should be filled with gravel and tamped for compaction.

5) *Protecting the trench*. Optionally, the gravel should be covered with a screed or flashing, to discourage water from reaching the gravel trench.

2.4.4.1.2. Substituting old poles in pole-and-trench foundations (experimental)

As said, the least durable elements of a pole-construction-based greenhouse are the poles. To substitute the foundation poles in place, the technique entailing the following steps may be useful.

1) Holes for new poles should be drilled away from the zone of influence of the existing poles, along the plane on which the poles lie. The choice of where to position the new poles depends on the arrangement of the edge beams and on the span between the old poles.

2) The new poles should be inserted into the new holes prepared for them.

3) The new poles should be connected to the existing perimeter beams.

4) If needed for durability, and if possible at the structural level, the existing poles should be cut in two, and the intermediate part should be removed, in order to disconnect the head of the poles from their foot with a complete capillary break. Optionally, the foot of the old poles should be removed by digging them away and refilling the hole.

5) Optionally, the heads of the old poles should be disconnected and removed from the edge girder.

2.5. Bracing strategies in timber frames

Several solutions that can be used for bracing light frames are less suitable for timber frames. Among them, there is bracing the structural bays with plywood panels. On the other hand, some strategies are not very suitable for light frames, but work well for timber frames, like that of sturdy bracing diagonals. The most important difference between timber frames and light frames, as far as bracing is concerned, is that the timber-frame elements, being sturdier, have more resistance to bending, and therefore can more easily resist the application of a force from a bracing element along their span (what is called an eccentric bracing – see Figure 2.104).

In general terms, in light frames, all the planes of the greenhouse structure should be braced, roof included, and in attached greenhouses, the building wall should act as a bracing for the rear part of the greenhouse, and should therefore be anchored to the greenhouse structure. Some common strategies for bracing timbers frames are presented in the following section. This information complements what has been said in the section on bracing strategies for light frames.

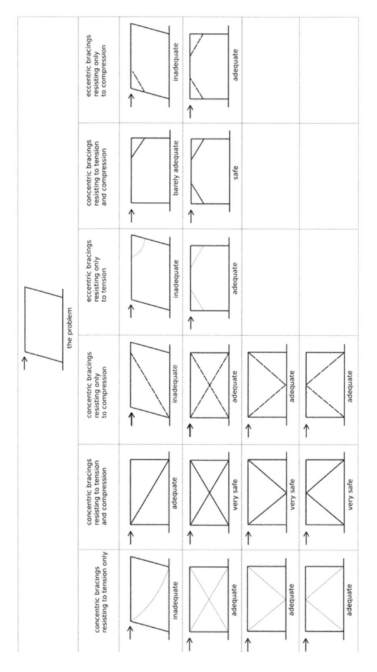

Figure 2.104. *Main possible structural bracing schemes for a single-bay bi-dimensional timber frame. Here, thin continuous lines signal diagonals only resistant to tension. Thick continuous lines signal diagonals resistant to tension and compression. Thick dashed lines signal diagonals resistant to compression only*

On references

The topic of frames is tightly linked to that of bracing. Some references about structural design, including calculations, are listed in Volume 4, section 2.4.1. For some introductory books not entailing calculation analyses, see Torroja (2011), Salvadori and Heller (1963) and Cowan and Wilson (1981). The book by Gauld (1994) adopts simple preliminary analysis criteria targeting a readership of architects. The book by Schodek and Bechthold (2013) is organized into three "layers" of increasing analytical depth, the first of which does not involve calculations.

2.5.1. *Bracing with cables or rods*

Bracing timber frames with steel cables or steel rods means that greater stresses are needed at the nodes than in light frames (Volume 4, Figure 3.36), and therefore requires stronger connectors to anchor the cables/rods, as well as stronger cables/rods. But the layouts and principles are similar in the two cases.

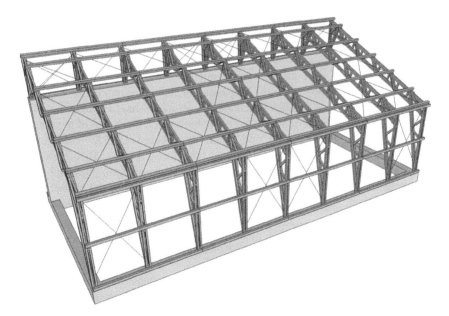

Figure 2.105. *The bracing action here is obtained with cables laid out in "X" configurations. The horizontal bracing serves the end and central bays between the uprights; the transversal roof bracing serves the end bays of the structure and the longitudinal roof bracing serves the bays at the perimeter*

2.5.2. Bracing with short massive diagonals

The possibility of bracing the frames with short, partial-bay-length massive diagonals exploits the lateral resistance to bending of the timber columns and the timber beams (see Figure 2.72). Today, the traditional timber-frame solution of diagonals connected with pegged tenons working mostly in compression (see Figure 2.26) is largely superseded by (a) the solution of the diagonals anchored to the beams and columns via butt joints executed by means of bolted, screwed or toothed steel plates (see Figure 2.34) and (b) the solution of lap-joining the diagonals with columns and beams by means of bolts or screws working in shear (see Figure 2.106). This happens both when the diagonals are short and when they are long.

2.5.3. Bracing the bays with full-length diagonals connected with butt joints by means of steel plates

In timber frames, lap joints for bracing the structural elements work less well than in light frames, because, in the former case, the timber elements used (diagonals included) are sturdier, the distance between the axial planes of the columns and beams, as well as of the diagonals, is likely to be greater. The greater sturdiness of the diagonals in timber frames often suggests connecting them, as seen, with butt-joints aided by bolted (or sometimes screwed or nailed) steel plates. It must be added that, because the full-length diagonals have a much stronger bracing capacity than the short ones, often bracing a few structural bays strongly is sufficient to make a whole timber-frame structure stable, both when the layout of the diagonals is multi-bay (see Figure 2.111) and when it is based on "X" configurations (St. Andrew's crosses) in rows of individual bays (see Figures 2.105, 2.106, 2.107 and 2.108). When the diagonals are short, partial-bay-length, they are usually distributed along with a greater number of structural bays, due to the greater bending stresses they put on the elements of the structure to which they are connected.

2.5.4. Bracing with full-length lap-joined diagonals

Because lap-joints are easier to execute than butt joints aided by cover-plates, and because they make the bolts (or screws, or nails) work properly into the wood (i.e. in shear rather than in tension), very often, timber-frame greenhouses are braced with diagonals connected by means of lap joints, screwed, nailed or bolted (like light frames), which often adopt some redundancy in the configurations, in order to guarantee robustness. This is a well-suited strategy for structures based on trussed light-framed portals, due to the distributed character of their connections.

The problem to solve, in those cases, is that of arranging the diagonals to avoid conflict with each other (because they overlap, or because they refer to the same connection space, anyway), both at mid-span and at their ends. The problem derives from the fact that, when two diagonals are concurrent in a place, if carefully chosen (and, often, bespoke) arrangements to avoid conflict are not taken, they should be reduced to half-wood configurations, which would increase the construction complexity and (more importantly) reduce the mechanical efficiency of the diagonals as bracing elements.

To avoid half-wood connections and conflicts, the diagonals can be arranged in non-symmetric, eccentric "X-patterned" configurations. Those configurations are less "reassuring" to the eye than the more usual symmetric and concentric ones, but when great hyperstaticity (deriving from redundancy) is given to the bracing structures, they do not entail an increased risk of instability.

Many of the examples in the following figures are of the asymmetric and eccentric kind. I'd like to underline that the reason why I am showing many examples of this sort is not that I think that arranging the bracings asymmetrically and eccentrically is preferable. This is to show that redundancy creates design freedom. We may or may not like these kinds of arrangements based on asymmetric and eccentric connections, but they work. This suggests that, in this kind of construction, apparent order is often overrated and control of the general underlying logic of solutions (of the kind that requires an understanding of the reason of things) is underrated.

The principal possible bracing arrangements among which a choice can be made are as follows. (a) The diagonals can be constituted by cables or one-bay-scale rods, suitable for working in tension and arranged as a series of X-configurations in rows across the greenhouse depth, and possibly also along its width (see Figure 2.105), or (b) they can be constituted by one-bay scale timbers, suitable for working both in tension and compression, with a similar layout as just seen – just, in the possibilities, less dense (i.e. involving a smaller number of structural bays), thanks to the greater rigidity of the solution (see Figures 2.106, 2.107 and 2.108). As an alternative, (c) the bracing function can be performed by a full-greenhouse-scale layout of cables or rods suitable for resisting in tension, arranged in whole-greenhouse-scale "X" configurations (rather than one-bay-scale "X" configurations) (see Figure 2.111), or (d) finally, it can be performed by a whole-greenhouse-scale open and continuous layout of timbers suitable for resisting in tension and compression. In this case, the arrangement of the diagonals can either follow a whole-scale "X" configuration or an open, non-symmetric configuration (renouncing to some strength and robustness) (see Figure 2.112).

When timber bracing diagonals in "X" configurations are used, both full-wood (see Figure 2.106) and half-wood connections are possible (see Figure 2.48).

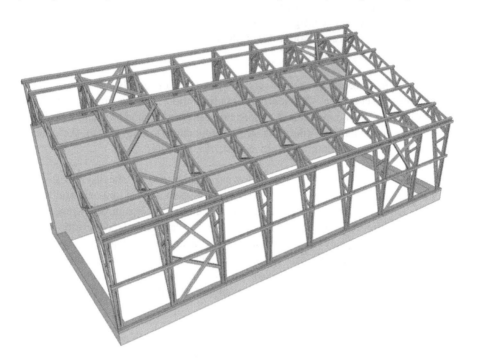

Figure 2.106. *This solution is braced with timber diagonals arranged in a markedly hyperstatic configuration*

NOTES ON FIGURE 2.106.– This bracing arrangement makes it evident that when X-configured bracing diagonals are lap-connected, it is usually not possible to make the bracing configuration symmetric and the bracing connections concentric. Indeed, it can be noted that the axes of the diagonals do not meet at the nodes. All the bracing arrangements shown here are indeed hybrids between concentric and eccentric. Concentric bracings do not need such a high hyperstaticity to produce satisfactory results, but are more difficult to construct. To avoid having to cut the diagonals in a half-wood fashion, some diagonals can be anchored to the purlins and others to the chords of the portal frames, in order to avoid putting them in conflict with each other. When the diagonals are concentric, instead, they: (a) have to be symmetric and (b) have to be anchored to either all of the purlins or all of the truss chords (see Figure 2.110). This, however, makes them enter into conflict with one another if they are not cut half-wood.

Furthermore, to make the bracing diagonal concentric, their axes have to meet at the nodes, which requires shortening their overlap in the lap joints at the nodes. Each of these diagonals, under a horizontal force (like, for example, that of the wind) works either in tension or in compression, in an alternated fashion. This bracing solution is even more hyperstatic than the one described previously, because its diagonals are laid out in an "X" configuration, rather than as a networked "chain" of diagonals. In this way, there is always one diagonal in tension and one compression in each structural bay, when shearing forces across each plane are present.

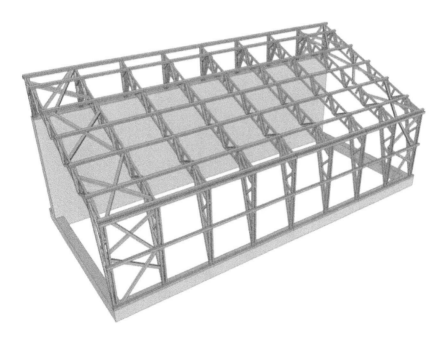

Figure 2.107. *Moving the X-braced bays towards the greenhouse ends makes the bracing more effective, provided that the braced zones do not become spaced too far away from each other*

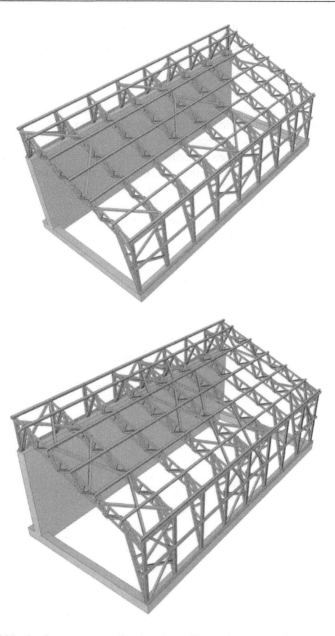

Figure 2.108. *In these cases, the bracing diagonals are positioned both at the extreme structural bays and at the central ones, for greater strength. The solution at the top is asymetric at the local level. The bracing of the solutions shown below is more complete, but conflicts between the diagonals in the intermediate part of the greenhouse can easily arise*

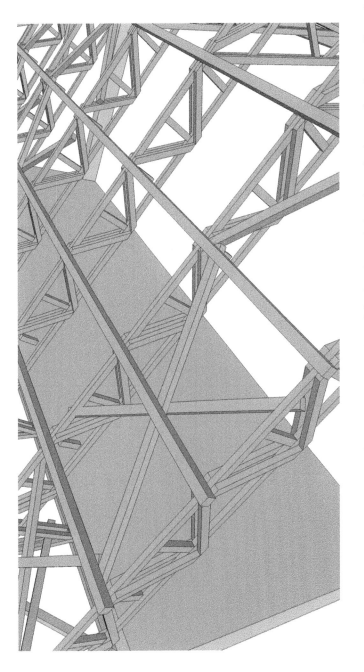

Figure 2.109. *This is image is a zoomed view on a structural bay of the bracing solution shown in the previous image. It can be seen that to avoid adopting half-wood joints in the middle of the bracing diagonals, one diagonal of each bracing pair has been connected to the purlins, and the other one has been connected to the upper chords of the beams. Therefore, the integrity of both bracing diagonals in each pair is guaranteed. The price to pay is that the solution is not symmetric, and is somewhat eccentric. The suitability of the bracing action here is substantially based on the redundancy of the solution, on the fact that it is highly hyperstatic*

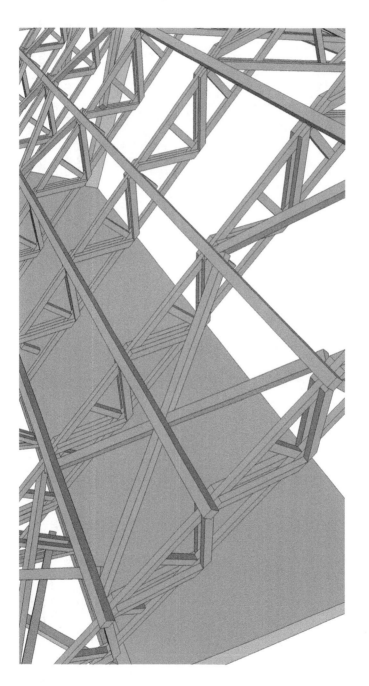

Figure 2.110. This view constitutes the symmetric version of the one shown in Figure 2.109. Symmetry is made possible by the fact that both diagonals in each concerned structural bay are connected to the purlins. The price is that the diagonals have to be arranged in half-wood configurations, which makes them substantially weaker in tension and compression. Note that, as an alternative, symmetry could have been obtained by connecting both diagonals to the upper beam chords

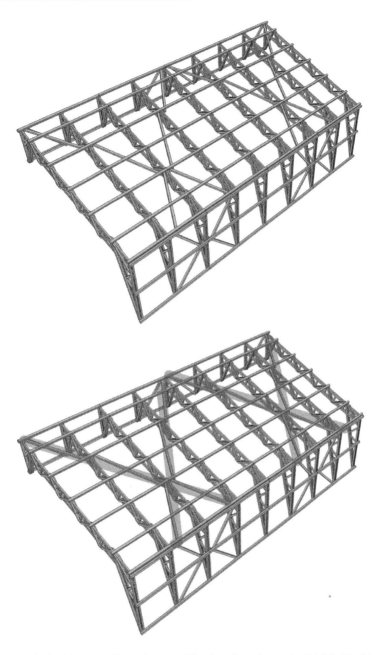

Figure 2.111. *In this case, the scheme of the bracing elements (highlighted in purple in the lower image) creates an X-configured "flow" of forces at the whole-frame scale, that is, at a scale greater than that of a bay: a flow that, in this case, has the scale of three bays*

NOTES ON FIGURE 2.111.– From this it can be seen that the bracing scheme can mostly be read at the scale of the whole greenhouse (see also Figure 2.108 (top)). This can happen despite the fact that the bracing elements are not continuous. Other bracing schemes would be possible based on longer and continuous, full-length bracing elements. This would require anchoring the diagonals to the upper chords of the truss portal frames, but the dense presence of diagonals in combination with the "density" of these specific truss beams would make the arrangement difficult at a constructional level. Overall symmetricity would certainly be more appropriate for a bracing scheme. But this bracing arrangement shows that redundancy can set the conditions for making asymmetry possible.

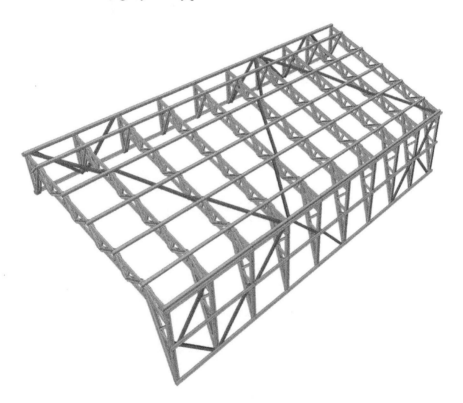

Figure 2.112. *In this case, the bracing elements (shown in purple) follow an open arrangement, which creates less conflict between the ends of the individual timber diagonals, but it is less symmetric and therefore less strong than the one shown in Figure 2.111. As shown in Figure 2.111, overall symmetricity would be more appropriate for a bracing scheme, but redundancy sets the conditions for making asymmetry possible*

3
Foundations

Foundations in greenhouses deal with gravity loads that are smaller than those dealt with in buildings, but the uplift and the horizontal thrusts they have to deal with due to winds can be considerable. This calls for some kind of specialization.

3.1. Foundation walls and foundation sills

The functions of a greenhouse foundation are to (a) support the posts of the frame and (b) to resist the uplift and lateral thrusts due to the winds. To perform the second function, the foundation must be sturdy or well implanted into the ground. However, in the case of provisional greenhouses, even the latter solution may be an overkill. In such cases, a foundation may indeed even be absent, and the wind uplift may be resisted by adding weights above onto the structure (or onto the roofs). For example, by means of sandbags (possibly hung) or bags filled with soil. But in all other cases, a well-anchored foundation is a good investment for the durability of a greenhouse and is therefore an essential part of it.

The foundation sill of a light-frame greenhouse can be built with masonry or concrete; however, under some circumstances, it may even be built with timber. Nevertheless, even in this case, it is not usually built as a framed solution but as a wall-like solution, because the connection between the foundation and the studs or mullions of the greenhouse is likely to statically form a hinge; and this in turn requires the supports of the greenhouse to have a certain resistance to lateral thrust and bending. This entails that, in a greenhouse, the foundation sill should usually be constructed as a continuation of the foundation wall. This fact also contributes to the necessity that the depth level at which the foundation is supported by the soil is usually rather superficial, due to the small load of the greenhouse itself.

For a color version of all figures in this book, see www.iste.co.uk/brunetti/greenhouses2.zip.

Because of their small weight with respect to their volume, greenhouses can exert substantial lateral and upward forces onto the foundation due to the wind. For this reason, the primary goal of a foundation wall is not only to support the vertical load of the greenhouse steadily, but also to ensure that the greenhouse does not move upwards or laterally. The conception of the foundation as a wall is very compatible with situations in which the greenhouse structure is distributed, as it happens in the light frames of houses. Dealing with a distributed structure makes the task of the foundation easier, but requires its anchor points to be frequent, which is hard to obtain with some foundation walls.

Whatever the solution adopted for the foundation is, the sill is usually constituted by an intermediate element of timber (pressure-treated lumber, larch or chestnut) used between the studs or mullions and the foundation.

In Figures 3.1–3.10, some kinds of foundation suited to greenhouses will be taken into account.

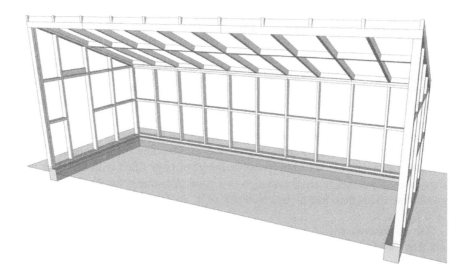

Figure 3.1. *The foundation most often runs along the perimeter of the greenhouse at a homogeneous depth. The position of the structure above the ground can be flush with the foundation, for the sake of constructive simplicity, and sometimes aesthetics, even if this makes the load on the foundation eccentric. This does not constitute a problem commonly because the loads involved are rather small (except when snow is allowed to dwell on a roof)*

Foundations 163

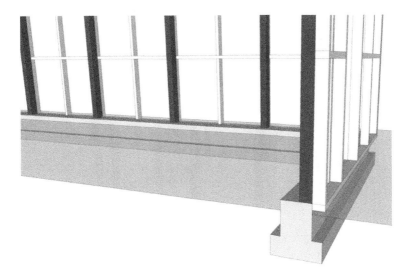

Figure 3.2. *Relation between a timber-frame greenhouse structure and a continuous foundation. The posts may be anchored to the foundation by means of steel connectors or a lumber plate (most often pressure-treated)*

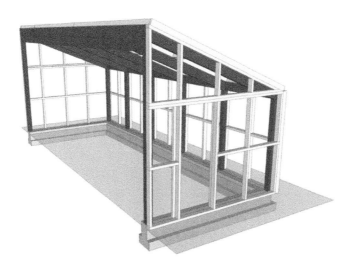

Figure 3.3. *The frames of the façades in this scheme of an attached greenhouse are anchored to a timber plate, which is in turn anchored to the strip foundation. It should be noted that the two timber posts in this example may be anchored to the shared wall. This may be done considering that the greenhouse structures without bracings have no rigidity of their own unless some bracing is integrated*

Figure 3.4. *In this greenhouse scheme for a lean-to greenhouse, thanks to the posts on the back of the frame, the structure is made completely independent of that of the shared wall. Indeed, this structure could even be suited to a free-standing greenhouse. The modification would only require adding the back façade to it*

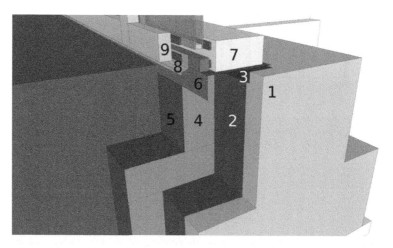

Figure 3.5. *Mullions/studs can be laid onto the foundation with a waterproof layer in between, or on a pressure-treated lumber plate with a waterproofing layer beneath*

NOTES ON FIGURE 3.5.– The front side of the foundation wall should be waterproofed with bitumen paint or other waterproofing paint, and with an impermeable layer; then, rigid, waterproofed and closed-celled insulating panels (most commonly, made of extruded polystyrene or glass foam) should be laid out onto the foundation walls and glued to them. Alternatively, a waterproof layer could be laid beyond the insulating panels rather than beneath them, or, as a further alternative, both beyond and beneath them. Another alternative entails the use of non-waterproof insulating panels; however, in this case, extra effort should be made, keeping the panels dry by means of well-drained gravel. In all cases, the insulation extends to "seek" the thermal break of the curtain wall at the transoms (more specifically between transoms and pressure caps). On the other side of the glass panel beneath the pressure cap, a batten of rigid foam insulation – for example, polyurethane – of thickness approximately equal to that of the glass is positioned above the insulation; and on top of it, an inner flashing is positioned, and a further external flashing is pressed in place by the pressure cap. The external flashing also has the purpose of hiding the insulation and the waterproof layer from view. The external soil will then be put back into the construction trench to press the insulation panels against the foundation wall. The level of the ground around the greenhouse may be sloped gently away from the greenhouse, in order to keep the foundation wall drier. Regarding the distance of the lower pressure cap from the ground, many sources advise keeping it at least 15 cm; but with all things considered, the length of this distance depends on how rainy a climate is. Certainly, the more rainy it is, the greater the preferred distance should be. A distance of 15 cm is suited to rainy cold climates, and this should therefore be sufficient for most other climates. Legend: 1. strip foundation wall; 2. waterproof mat (here or at position 4); 3. waterproof mat; 4. closed-cell insulation; 5. waterproofing layer (here or at position 2); 6. flashing; 7. transom (possibly resting on a moist-resistant timber plate); 8. flashing; 9 pressure cap.

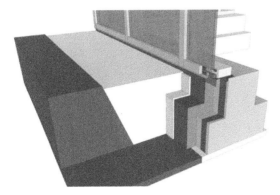

Figure 3.6. *The foundation trench may be filled with gravel to keep the zone of the foundation drier. Also, the external face of the insulation panels, or the waterproof layer outside the panels, may be lined externally with a drainage blanket, which is usually a plastic geotextile*

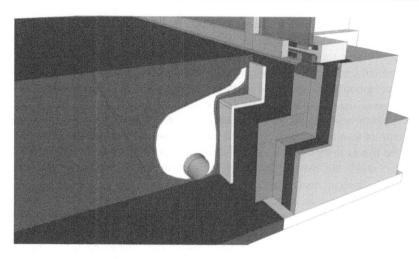

Figure 3.7. *The most complete solution entails positioning a filtering blanket in the foundation trench, putting gravel on it and then a perforated pipe; then filling the trench by half with gravel; then wrapping the gravel with the filtering blanket; then filling the remaining part of the trench with earth, and tamping it, in order to compress it. The filtering blanket prevents the gravel from clogging the soil; and pressing the earth above discourages the water from seeping through the trench towards the foundation wall*

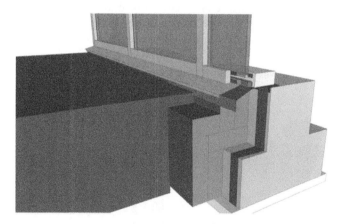

Figure 3.8. *Insulating the foundation wall as much as the greenhouse is economically worth it only if the greenhouse is attached to a building and it has to be as inhabitable as the building. The insulation sizing in this example is typical of a solar greenhouse attached to a passive-house-standard-level building*

Foundations   167

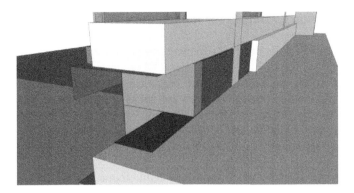

Figure 3.9. *The bottom transoms of the façade can be heightened with respect to the top horizontal plane of the foundation wall, in order to make room for the insulation beneath. This is usually worth the trouble economically only if the greenhouse is inhabitable. It should also be noted that, in this case, the mullions are supported by the foundation wall directly, and not by the transoms which are discontinuous*

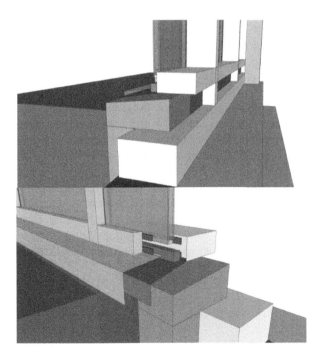

Figure 3.10. *A frequent alternative to the previous solution is to place a pressure-treated timber plate below the horizontal insulation and make it work as a mediation layer between the mullions and the foundation wall. In this case, the insulation panels rise above the ground level and therefore should be protected by some kind of sill – the simplest of which is constituted by flashing*

3.2. Construction strategies for foundation walls

The foundation walls of a light-frame greenhouse can be built in several ways, provided that they respond to some fundamental requisites: (a) the ability to support a light and mildly concentrated load exerted by the studs, and to resist the everlasting and random thrusts and pulls exerted by the anchors under the effect of the wind; (b) the ability to separate the soil within the greenhouse – or the greenhouse floor – from the soil outside the greenhouse, providing an even support to the thermal insulation, the water-impermeable barrier and the vapor barrier.

Because the foundation walls of a light frame are loaded in a rather homogeneous way, they do not require high bending resistance. Therefore, when the foundation walls are made of concrete, the concrete may be only mildly reinforced, or not reinforced at all, and when they are made of masonry, the masonry can be of many possible kinds.

In any case, the construction of the foundation walls often takes place after the piping (drains, water lines – when present) coming from, and going to, the greenhouse, taking care to avoid laying them down too near to the electric lines. The water lines should be positioned below the frost level of the ground. This is something that should also regard the foundation support plane: this too should be located below the frost level.

In temperate and temperate-cold climates, the depth of the foundation of glazed greenhouses should usually be between 30 cm and 1 m, or deeper, depending on the freezing depth of the ground. This is because the bottom of the foundations should reach a depth below the possible freezing level of the ground, so as to avoid the upward thrust that the soil exerts when it freezes due to expansion[1]. The most superficial depths (30-50 cm) are by far the most common, because very often it is accepted (and acceptable) that the ground below the foundation can freeze (and expand). In the case of glazed greenhouses, this can happen when the greenhouse is conceived with a capacity to adapt to movements without propagating the stresses which can be dangerous for the glass panels. This is especially important in attached greenhouses.

In zones where the ground will not freeze, the foundations are not mandatory at the thermal level, but it is still very advantageous at the construction level. In such cases, the foundation can be superficial.

The foundation walls should be waterproofed externally, and the waterproofing may be used for gluing the insulation panels. Also, the soil around the foundation

---

1 The expansion of the soil when it freezes is, of course, due to the ice being less dense than water, and therefore it occupies less space – the quantity of material being equal.

wall, externally to the greenhouse, should be drained effectively by creating a trench and sloping it away from the greenhouse, by filling it with gravel, as well as by possibly placing a perforated drain of about 10 cm in diameter in it. The obtained French drain may then be sloped towards a dry well – that is, a well, or a trench, filled with gravel, conceived to maximize rainwater absorption.

3.2.1. Preparing the ground for a foundation wall

When the top soil is unsuited to supporting a foundation wall, whatever the material of the foundation wall is, a trench should be dug in the ground for the hosting of the foundation wall itself. The absolute minimum depth of the trench is 20 cm, but at least 30 cm is recommendable, with a preference for 40 cm. When the weight of a greenhouse is light, the height of the foundation trench is more dependent on the strength and rigidity of the foundation itself than on its weight. For this reason, the weaker and more incoherent the foundation wall is, the deeper the foundation trench should be. The greatest depth (and width) would be required for a foundation wall built with stones without mortar: in this case, it would require 60 cm of depth below the ground. The smallest depth would be required by a reinforced concrete wall, for which a depth of 20 cm below ground may indeed be enough to support a small greenhouse.

In each case, the ground supporting a foundation wall should be made level (checking its horizontality with the water bubble level), then finished with a layer of 5 cm of lean concrete (best solution) or gravel, in order to make the support even and keep the reinforcing bars of the concrete separated from the ground.

3.2.2. Boulders-and-mortar wall foundation

The boulders-and-mortar wall foundation is the hardest type of foundation to build and the weakest type mechanically, and for this reason, in this book, how to build it in a facilitated way, using formworks, will be taken into account.

What is required are lateral formworks as in the case of a concrete foundation, braced against the lateral thrusts and tied to each other. Then, it is necessary to lay out, stone by stone, on an abundant layer of mortar, a layer of masonry, without leaving large voids in the blockwork, and filling the voids with mortar. Another layer of cement mortar should then be laid on the stones, and then other stones should be laid out, in an alternated fashion, and so on, until the due height level is reached. Finally, the topping of the wall should be completed with a layer of mortar about 4 cm thick, leveled horizontally.

Due to the low coherence and to the irregularity of the described foundation wall, it is difficult to firmly anchor a timber plate (sill) on top of it. If expansion dowels are used for anchoring it, there should most likely be many, in order to lower the stress to which they are submitted. An alternative way to anchor a greenhouse sill on a foundation like this is to embed bent rods into the blockwork during its construction. Each anchor rod should preferably have a threaded end, in order to make it possible to bolt it to the plate. As an alternative, the ends of the rods may be bent with a hammer, in order to hold down the timber plate.

3.2.3. Brick masonry wall foundation

When fired clay bricks are used to build the foundation walls, the trench into which they are placed should afterwards be filled with gravel, in order to make water drainage possible. This is because fired clay bricks are porous and absorb water. The brick wall should be at least three brick heads thick. If it is narrower, it may lack the strength and the weight for resisting uplifts and lateral forces due to the wind. This is true unless the greenhouse is very small. In this case, the foundation wall may be made two bricks thick.

3.2.4. Conventionally mortared hollow concrete block wall foundation

A concrete block wall presents the advantage that some of its cavities may be filled with concrete, in order to make the wall stronger, and possibly with steel rods, of the type used in reinforced concrete (see Figure 3.12). In this case, the blocks do not necessarily need to be mortared all along their thickness; they may be just mortared in strips near the inside and the outside face, vertically and horizontally. In the case of ribbed hollow concrete blocks, this is not just an option, but the only solution available.

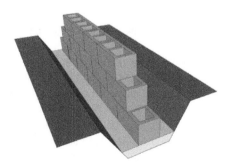

Figure 3.11. *Hollow concrete (sand-and-cement) block walls are the quickest alternative to reinforced concrete strips foundation walls*

NOTES ON FIGURE 3.11.– This wall can be easily reinforced into some of the vertical cavities where bars can be placed and concrete can be poured; however, they do not have the horizontal bonding of reinforced concrete foundations. Horizontal reinforcements can be integrated into these block walls, in the horizontal joints between the blocks, but in this position, the steel rods are not very protected from corrosion (unless they are made of stainless steel) and do not collaborate with the blocks as thoroughly as they would in concrete. Despite these shortcomings, the obtained resistance is usually more than enough for a greenhouse.

When the cavities are not filled with concrete, they should at least be closed with mortar at the top, which may be obtained by closing them provisionally with something like crumpled newspapers (Jones and McFarland 1984), so as to create a support for laying out the mortar. A concrete pour for every two vertical cavities produces a very sturdy wall, and a pour for every cavity makes it even stronger and watertight.

As anchors for the timber sill plate, even in this case, threaded steel rods may be used as anchor rods; or L-shaped steel rods (of the kind for reinforced concrete) may be bent to hold the wooden plate down.

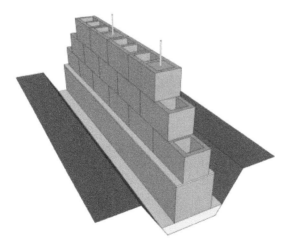

Figure 3.12. *The concrete foundation and the hollow block masonry foundation can be compounded into one by sitting the block wall on top of the reinforced concrete footing strip. This is especially useful when, for some reason, a tall (relatively speaking) foundation wall is needed: for example, when the basement of a greenhouse has to be filled with gravel, or has to host a hollow basement for convective heat transfer*

3.2.5. Parged hollow concrete block wall foundations

An alternative technique for building a foundation wall with hollow concrete blocks quickly is to lay out the blocks without mortar and then bond them together by putting a layer of special mortar constituted by a mix of sand, cement and fiberglass onto both sides of the wall. This solution, called parging, is an even quicker alternative to the ordinary concrete block wall foundation, which, on the other hand, is usually stronger.

3.2.6. Concrete foundation walls

A concrete foundation wall is a regular inverted beam-like foundation reinforced with steel rods. The anchors can be embedded in concrete along with the reinforcement bars. This is by far the strongest and most watertight among the foundation walls (see Figure 3.13).

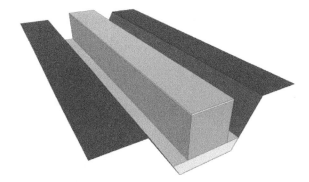

Figure 3.13. *The shallowest and most practical greenhouse foundation is constituted by a mildly reinforced concrete strip*

3.2.7. Wooden-frame foundations

Pure wooden-frame foundations, constituted by a wooden plate on gravel anchored to the ground by means of stakes, and surmounted by light frames, are rare but possible. Their main problem is of course durability, because the ground is a moist environment and their section is small. For this reason, these foundations tend to be used in non-permanent situations. Their construction resembles somewhat that of timber log foundations (presented in section 3.2.8.1), but is lighter. The fact that these foundations are laid out on gravel contributes to keeping the underground environment dry and extends their durability. A much more common variant of light-frame foundation is constituted by joists anchored to the ground by means of stakes (see section 3.2.7.1).

3.2.7.1. Stake-based foundations

A light and reversible type of timber foundation (a) can use steel stakes that are hammer driven (or mallet driven) into the ground, and anchored via screws to ground joists, which in turn support the greenhouse studs/mullions (see Figure 3.14, and Figure 5.1 in Volume 4), or is directly anchored via screws to the foot of each stud/mullion (see Figure 3.15); or (b) can use wooden stakes of pressure-treated lumber or chestnut lumber hammered (or malleted) into the ground, to which the ground joists are anchored (see Figure 3.15).

A variant of steel stakes is constituted by steel anchor screws that are driven into the ground by rotation. They perform a work similar to that of steel stakes, but offer a stronger resistance to uplifts.

In all these cases, the closure against the ground is performed by the ground joists (see Figures 3.14-left and 3.16) or by additional edge joists (see Figures 3.14-right and 3.15-right) – all of which should therefore be of a rot-resistant lumber, like chestnut lumber.

The advantages of this kind of foundation are its lightness and reversibility.

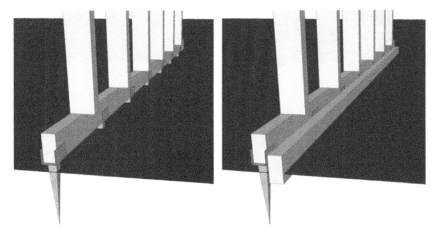

Figure 3.14. *Scheme of a light-frame foundation based on ground joists anchored to steel stakes. The stud/mullions are screwed to the ground joist. The side ground joist (right), screwed to the main ground joist, works as a "skirt" separating the soil of the greenhouse from the external soil*

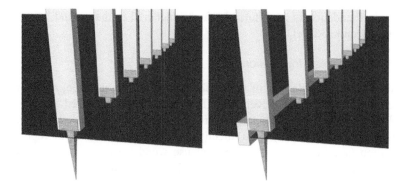

Figure 3.15. *Scheme of a light-frame foundation based on stud/mullions anchored to steel stakes. The stud/mullions are screwed to the steel stakes. The side ground joist (right), screwed to the joists, works as a skirt separating the soil of the greenhouse from the external soil*

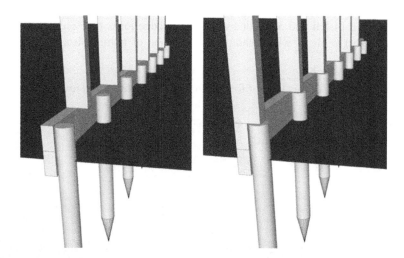

Figure 3.16. *Scheme of a light-frame foundation based on edge joists anchored to wooden stakes. In the left and right images, the studs/mullions are placed in different positions with respect to the stakes, but, in both cases, they are screwed to the ground joist, and the ground joist is screwed to the stakes. The load-bearing ground joist is the upper one, while the lower one works as a skirt separating the soil of the greenhouse from the external soil*

3.2.8. *Timber foundations*

The most common log foundation, and the only one that will be analyzed here, is the timber log and gravel trench foundation, which is kept dry by the presence of the gravel. An authoritative precedent of this type of foundation was used by Frank Lloyd Wright in some houses of his early period (and described in print – Wright 1954), as well as in some of his later wooden Usonian houses.

3.2.8.1. *Timber log and gravel trench foundation (experimental)*

In this section, an experimental strip foundation solution adopting a "dry", mortarless technique is presented. It is a low-cost type of foundation that can be built quickly and easily, as experimented with in the case study of the greenhouse shown in Volume 4, section 3.2.

In this technique, a timber log is placed onto a gravel bed in a trench, making drainage possible, as well as making it possible to use untreated lumber as a durable foundation material. The timber logs forming the top part of the foundation are anchored to the ground below by means of steel rods (of the kind used for reinforcing concrete) driven through the gravel into the underlying soil diagonally.

The construction procedure entails the following steps:

1) As a first step, timber logs suitable for supporting a house in moist conditions have to be found. They have to be large enough, strong enough, low-cost and, most importantly, durable enough for this kind of usage. Here, the main point to consider is that the durability of common timbers is rather low when they are embedded in the ground. The gravel used in the described scheme has indeed the function of draining water away from the wood, without reducing load-bearing capacity of the soil too much.

There is more than one strategy for choosing timbers for a timber-and-gravel trench foundation. The first strategy is to choose intrinsically durable timber species. Larch and chestnut are the most suited species. The second strategy is to choose ordinary timber species with improved durability. A solution for improving the durability of ordinary construction lumber is pressure treating it with preservatives in production. The drawback of this strategy is that preservatives, which have been observed to be effective in killing wood parasites and preventing the decomposition of timber caused by microorganisms and fungi, are usually not good for human health. The fact that preservatives are likely to be poisonous to the soil is always a bad thing, and it is even more so when the soil is used for growing edible plants. A third preservation strategy, which has been used since ancient times, is charring the external surface of timber – especially the end of the posts to be buried underground – by treating it with fire. Through this treatment, the external part of timbers becomes mineralized, which increases their durability and resistance to rot in moist conditions.

To burn the surface of timber properly, a fire may be prepared. A quicker solution suited to today's methods is to use a kind of flame suitable for soldering the waterproof layer of roofs. Only the surface of the wood has to be burned in this process; otherwise, the timbers may lose a substantially useful structural mass, as well as a substantial part of their mechanical strength.

The described strategy has evolved into the industrialized one of heat-treating lumber. Among the types of heat-treated lumber, there is baked or roasted timber (timber heated for hours in an oven), which is an effective solution for mineralizing ordinary timber and making it more resistant to decay. However, roasted lumber is currently still too expensive to be used for foundations.

2) A continuous trench has to be dug down to a depth suited to bearing reinforced concrete foundation walls equivalent to the foundations to be built. This depth depends on the soil condition and the load of the structure to be built. The common depth of the trench may be of 40 or 30 cm. The width of the trench should be somewhat greater than the width of the timbers that are used – about 20–30 cm more by each side. The bottom of the trench should be as flat as possible and well compacted, and the side walls of the trench should be as vertical and straight as possible, as allowed by the soil. This will assure that the loads are spread evenly only at the bottom of the trench, rather than at the sides, in order to decrease the likelihood of gradual settlement of the foundation under load.

3) The sides of the trench should be covered with a filtering cloth before filling them with gravel. Enveloping the gravel with cloth prevents the gravel from getting filled and clobbered with drained earth over time.

4) Ideally, the gravel that is used to fill the trench should be composed of elements greater than 3 cm in diameter, of even size and round shape. This makes it likely that the voids between the grains are wide enough to break the capillary force in the presence of water and make the gravel well suited to draining the water. The trench depth must be sufficient for bearing the load of the greenhouse that is going to rest on it; similarly, the height of the gravel must be sufficient to bring the foundation timbers beyond the level at which they can get soaked by water during the rainy season. If the height of the trench is substantial – say, more than 30 cm – the lower layer of gravel may be substituted with layers of stone pebbles, or crushed rocks, to the advantage of load-bearing capacity and cost containment.

5) The layers of gravel, or stones and gravel, should be compacted well by pounding them with mallets or the flat heads of carpenters' hammers.

6) The timbers chosen for building the foundations should be put in place and aligned in position, possibly in such a way that the loads transferred to them are applied at or near their central axis, in order to avoid subjecting them to torque. This would occur if the loads are applied at locations too distant from the central axis of the foundation.

7) After positioning the timbers, they should be anchored to the ground by means of steel rods. To do this, professional ground anchors can be anchored to the timbers with bolts, and strong and large washers can be used to avoid that, under tension, the heads of the bolts make their way through the timbers. As an alternative, anchoring may be obtained using steel rods bent so as to hold the timber elements at the foundations (see Figure 3.20). In this case, each steel rod should be bent at 90° at its top end, in order to obtain a horizontal chunk long enough to hold down the foundations when the winds pull it up.

The length required for this top end will mainly depend on the strength of the timbers used for building the foundations (with the principle that the stronger they are, the shorter the segment will be) and their width. A length of about 10 cm of rod heads should be appropriate in most cases.

The rods, or ground anchors, may be placed in holes prepared in advance in the timber logs, or the rods may enter the ground at the sides of the foundation logs. In both cases, the rods should be slanted with angles opposite to those of the rods nearby, so as to make the obtained anchorage work against the forces pulling the foundations upwards, regardless of the direction and verse of the horizontal component. If the ground anchors traverse the logs through holes, the holes should be positioned on the basis of a scheme of the kind shown in the plan and section in Figure 3.21-e, f, g, or of the kind shown in Figure 3.21-k, l, which increases the volume of ground reached by the rods shown in Figure 3.21-m, n, o.

Each timber log in these solutions is held down by the heads of the rods that have been hammered into the ground for keeping the logs in place firmly.

The holes in the timber logs may be prepared in advance offsite, because preparing them on-site would be more difficult. If one of the solutions with slanted rods is adopted, it is necessary to check that the tilt of the rods is not greater than what is allowed by the available space around the trench. The rods should be driven into the ground at a depth sufficient to hold down the foundation, which implies that they should rest in the gravel for no more than one-half of their length. In addition, they should be driven into the soil for at least 40 cm. This is because the portion of the rods that traverses the gravel does not add much to the resistance to the extraction of the foundation, differently from the portion of the rods that enters the ground.

In cases in which the logs are not prepared in advance with equally spaced holes, they should be held down by placing the rods (and their heads) in a zig-zag layout, as shown in the scheme in Figure 3.21-e, f, g, with a longitudinal slant.

A transversal slant is more difficult to obtain because it should be accompanied by an additional bend at the rod heads, but it can be implemented with small effort. The slant of the rods may be directed towards the inside or the outside (see Figure 3.21-k, l) of the greenhouse.

8) After hammering the anchors in place, the rod heads should be hammered down to bend further and hold the logs as firmly as possible. Then, the voids between the logs and the trench walls should be filled with gravel, and the gravel should be compacted – as observed, by pounding it with hammers or poles, in order to make it hold the foundation as firmly as possible in its place (see Figure 3.19).

9) A waterproof mat or, in the lowest-cost case, a barrier constituted by a polyethylene film (ideally overlapped in a double or triple layer) should be placed on the foundation logs, in order to avoid water being sucked (due to the capillary force) into the timber plates that will be screwed or nailed onto them (see Figure 3.20-h). Finally, the bottom plates of the greenhouse frames can be nailed to the foundation timbers, with a spacing no longer than 1 m (see Figure 3.20-i).

a)   b)

Figure 3.17. *A dry, "quick" type of continuous foundation can be laid out on gravel. The trick is how to make this foundation suited to a greenhouse, which has to resist the wind's uplift*

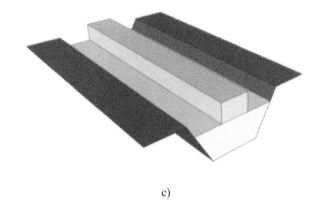

c)

Figure 3.18. *The first construction step after the trench is dug and the gravel is placed to fill it almost completely is to place the logs of suitable timber type on top of the gravel, and pounding them to make them horizontal*

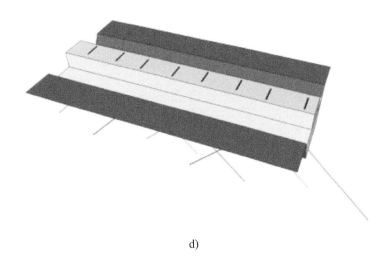

d)

Figure 3.19. *The next step is to drill holes into each foundation timber log at regular intervals, in a zig-zag fashion. Then, reinforcing rods should be cut to size, and their head should be bent 90° by one side (to hold down the timbers when hammering the rod head into them until the end-run). Each rod should reach the ground, passing through the holes in the timber logs, diagonally and in a mutually divergent way, so as to give resistance to the extraction of the foundation. The rods should be long enough to penetrate the ground layer below the gravel and stick into it*

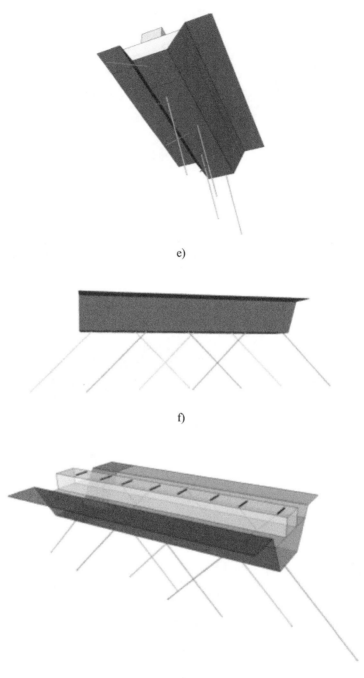

e)

f)

g)

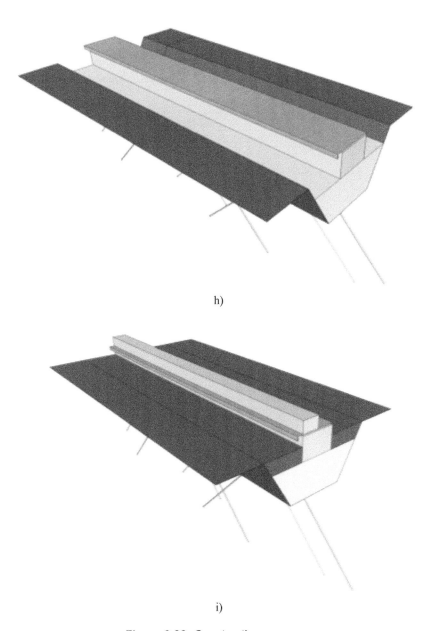

h)

i)

Figure 3.20. *Construction sequence for the gravel trench foundation*

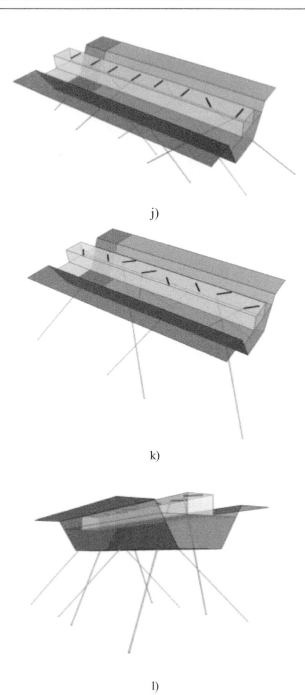

j)

k)

l)

Foundations 183

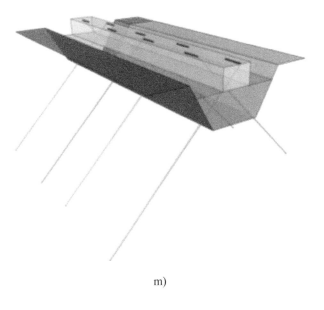

m)

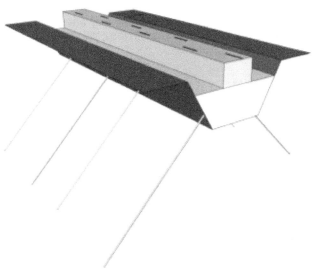

n)

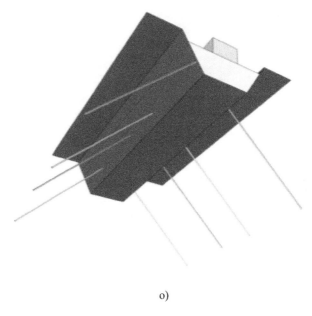

o)

Figure 3.21. *Several alternative arrangements can be given to the rods driven into the ground, provided they are divergent and distribute the stresses into the ground evenly*

3.2.9. *Pier foundations*

Pier foundations are part of a structural frame that raises the level of the beams or girders above the ground. The girder anchored to the piers can be of the same materials as the piers, when those are of steel, reinforced concrete or timber, or it can be a timber one, independently of the material of the piers. The girder supported by the piers performs the function of a foundation wall, and it is related to the structure in the same way. The described solution is usually adopted when the greenhouse floor is going to be above the ground level. This is due to the difficulty of closing the space between the lower beam and the zone of the foundation in a durable, performant and simple way. In any case, with a light-frame greenhouse, a timber plate is usually present between the girder and the studs/mullions.

3.2.10. *Insulation of the foundation wall*

The quantity of the insulation in a foundation wall depends on the level of thermal efficiency that the greenhouse is expected to have. The higher the efficiency, the thicker and deeper the insulation of the foundation wall should be.

The following gives a broad idea of the performances involved: for a low-to-average thermal efficiency, the U value of 0.5 Wh/m²K may be chosen; for a low thermal efficiency, no thermal insulation at the foundations are needed; and for a high thermal efficiency (superinsulation), a U value of about 0.2 Wh/m²K may be chosen. The latter value may be chosen for an inhabitable greenhouse aiming to the passive house energy standards.

The insulation material used should be adequate for producing rigid panels that, being closed-celled, are capable of maintaining their thermal resistance when they are also wet or moist: for example, as in the case of extruded polystyrenes (synthetic), closed-celled polyurethanes (synthetic) or foam glass panels (mineral). But even the use of partially non-closed celled materials such as fiberglass panels or perlite panels are possible when the foundations are kept sufficiently dry. This can be obtained by filling the foundation trench (of course, external to the wall) with drainage gravel – and possibly positioning a gently sloped perforated pipe at the bottom of the gravel, in a French drain configuration.

The insulation panels should be glued onto the support constituted by the foundation wall, and at the same time, they may be anchored to it with expansion dowels. Regarding the position of the insulation, the principal choices are outside or inside the foundation wall. In all cases, the position of the insulation should be arranged so as to aim at the wooden frame, or at its thermal break, if present (see Figures 3.5 and 3.10).

Insulation in the outer position is more suited to making the thermal mass of the foundation thermally useful for the greenhouse; and insulation in the inner position is more suited to obtaining more thermal responsivity, which can be useful in the case of heating and air conditioning based on all-air mechanical plants. As regards the depth of the insulation, construction-wise, the appropriate choice is to accompany the foundation wall all the way down with it, down to the plane of structural support, but not deeper than that, because this occurrence would weaken the foundation, by subtracting useful mechanical resistance from the ground beneath it. However, this structural consideration conflicts with thermal requirements, which in theory would advise bringing the insulation to a deeper level – optimally, at least at a depth of 3 m, in ultra-performant, highly passive, residential solutions. A solution to this conflict is to extend the insulation horizontally rather than vertically, from the plane of support of the foundation – ideally for a total length (vertical portion + horizontal portion) of about 3 m in ultra-performant situations. In this arrangement, the insulation forms an umbrella-shaped thermal barrier below the building that has the effect of making the thermal bridges long enough to make them practically broken (see Volume 4, Figure 1.41). The result is not equivalent to when the mass is insulated by insulation arranged vertically in depth, because it involves more mass and that mass is in more diffuse contact with the surrounding soil; however, it is the best solution obtained with an ordinary foundation configuration.

3.2.11. *The foundation wall as a sill*

The foundation sill (the top of the foundation wall) should usually be protected by flashing, and a certain slope towards the outside should be given to this flashing, in order to avoid water remaining on it. In addition, this flashing should usually "cantilever" inwards and/or outwards, in order to protect the insulation layer applied onto the foundation wall (see Figure 3.5).

3.3. Drainage around the foundation wall

Drainage around the foundation wall is important to avoid letting the ground around the greenhouse get soaked. Drainage has great advantages in reducing the risk of movements of the ground near the foundation due to soaking or freezing below ground; as well as regards durability, especially with respect to low-permeability soils such as wet clayey soils.

The most efficient strategy for assuring good drainage is to create the aforementioned "French drain": a trench below ground level filled with large-grained round gravel, to reduce capillary forces; it would be better still to put a sloped perforated pipe to collect water into it, at its bottom. Especially in residential interventions, the drain pipe may then run, with a slope, towards the part of the ground where a system suited to increase the ability of the ground itself to absorb water is built – a gravel well or trench. For example, a trench or well filled with gravel, so as to increase the capability of that portion of ground to absorb water. A suitably shaped flashing should bring rainwater into the trench. As an alternative, water may even be conducted there by prolonging the polyethylene films into the trench, making them perform both as films and underground gutters (see Figure 3.7). However, this solution has its drawbacks: the channeling arrangement of the water into the trench will have to be refinished each time the films are changed.

To keep the ground around the greenhouse as dry as possible, it is also useful to face the ground surface slope gently away from the greenhouse, when possible. But this does not eliminate the need to waterproof the foundations externally, in order to avoid water seeping into the greenhouse from the outside.

3.4. Pavements

Pavements in the areas of a greenhouse not left to the growing soil can be designed to maximize heat storage and/or water drainage. Maximizing heat storage is especially important in solar greenhouses, while water drainage from the pavements has to be pursued in all kinds of greenhouses.

It has been observed that to pursue thermal storage, it is necessary to make the pavement conductive and locate the thermal mass below it, in contact with it. A typical example of construction configuration attaining this is a concrete slab with a pavement of fired clay tiles on top, laid out on mortar. If the pavement were laid out on sand (which is difficult to do in practice unless the tiles are very thick, or bricks or concrete locking blocks are used, rather than tiles), it would reduce the thermal conduction to the slab and the efficiency of thermal storage. A similar consequence would be expected when casting the concrete slab on gravel or crushed stones, rather than on compressed earth. The effective thermal capacity of the system constituted by pavement+underfloor+slab would certainly be greater than that constituted by the only pavement. However, because of the sand or gravel, that system would not use the thermal capacity of the ground fully.

In a solar greenhouse, laying out the pavement+concrete slabs directly on the compressed soil is preferable at a thermal level, even when positioning the slab onto a gravel bed – in order to favor drainage – would be preferable at a construction level. An inconvenience of the described scheme is that the drainage has to be created on top of the floor, which therefore should be sloped conveniently (which may be a nuisance from the viewpoint of usability) – that is, with a slope of the order of 1%, towards a water collection point or towards the soil, depending on the layout and aim of the greenhouse.

The drainage points of the greenhouse may be one or more (depending on the greenhouse size). If the point is only one, it is usually at the center of the greenhouse. As an alternative, it may be located near its perimeter, especially when the greenhouse is small (Shapiro 1984; Marshall 2006; Schiller and Plincke 2016).

Each drainage point should be connected with a drainage pipe exiting the greenhouse through the foundation wall, set in position before building the foundation wall.

An alternative solution to the sloped impermeable floor, suited to obtaining efficient drainage (although at the expense of the thermal mass) and a complete horizontality of the pavement, is constituted by using (a) a brick pavement (in this case the bricks should be laid flat on the side or on the header, depending on the quantity of mass pursued) on top of (b) a layer of 10 or 15 cm of sand, or a layer of about 5 cm of sand on top of 10 or 15 cm of gravel, separated by a geotextile layer (which, in the simplest case, may be constituted by a thick polyethylene film) – in order to prevent the sand from clogging the gravel – or a layer of 15 cm of fine gravel. In its turn, this layer should be laid on top of (c) a gently sloped concrete screed, (d) a concrete slab and (e) a screed separating the slab from the ground. The slab and the screed above may be cast as one – that is, the pitch may be given to the slab directly.

A solution characterized by a light pavement detached from its support – for example, wooden planks on battens on a sloped (for drainage) concrete slab, or a pavement of planks on battens laid out on gravel (and stacked to the soil) – due to its lack of effective thermal mass is suited to the cases in which the greenhouse is used as an air solar collector. In this case, the ability to reach high air temperatures during the day is accompanied by the symmetrical tendency to reach low temperatures during winter nights.

3.5. Platform frame floors raised above the ground

Platform frame floors may also be supported by short poles and kept above ground (see Figures 2.89, 2.90 and 2.93–2.100). This is a solution that may be chosen, for example, when the soil is not suited for growing crops. In this case, when plants are present in the greenhouse, they should be cultivated in pots sitting on the floor or benches.

When large spans are sought, the construction of the floor may be most commonly:

a) Timber-framed or light-framed, with primary beams orthogonal to the front façade, or parallel to it, with or without secondary beams orthogonal to the primary ones, and with the joist sitting on the primary beams or, when present, on the secondary beams, and with plywood or OSB sheets as an enclosure, above and below the joists. When large spans are not sought, the floor may be light-framed In this case, the floor may be a suspended platform-framed one anchored to a double header joist (ring beam) between the ground poles, or sitting on top of them (see Figures 2.93, 2.94, 2.100 and 2.101).

In any case, in a growing greenhouse, one buttonhole should be positioned at each extreme between each joist at the header beams or at the intrados of the floor, in order to let the water vapor out of the floor cavity and improve on the result of the vapor barrier. As always, in climates ranging from temperate to cold, this must be positioned on the warm side of the enclosure: in the case in question, above the structural part of the floor, just below the primary sheeting on top of the joist, or just above it, or below the pavement system.

4

Heating and Cooling Systems; Watering Systems

4.1. Heating and cooling plants

Heating and cooling systems for greenhouses should be chosen and sized so as to supply the heat or coolth that passive solar methods cannot provide in certain climatic situations. Therefore, they should be employed when all other possible measures to reduce the energy load of a greenhouse have been taken. In the following sections, the most common solutions for greenhouse heating and cooling are reviewed.

On references

Grondzik et al. (2019) is a major reference on mechanical systems in architecture. Albright et al. (1978–2001) is a reference about greenhouses, particularly strong in mechanical systems integration. In Esen and Yuksel (2013), the effects of several renewable sources for greenhouse heating are compared experimentally. Ouazzani Chahidi et al. (2021) analyses the behavior of a plant-integrating greenhouse in a Mediterranean climate.

4.1.1. *Air-based systems*

The cheapest strategy for heating and cooling a greenhouse actively is to use a centralized air system. The limitations of that kind of system derive from the fact that, because air has little thermal inertia, it is not very good at influencing the

---

For a color version of all figures in this book, see www.iste.co.uk/brunetti/greenhouses2.zip.

temperature of the thermal masses of a greenhouse. The low heat capacity of air also entails that the radiant temperatures in the greenhouse's indoor environment will respond with a substantial delay, and inefficiently, to the "injections" of heat and coolth, when they will be done via the air.

The cheapest way for heating air actively is by means of a boiler, and the cheapest way to deliver that heat is by means of fan coils served by water pipes connected to the boiler, and powered (as regards the fans) by electricity. The main alternative involves a heat pump (or more than one, for example, when the presence of a backup in case of failure is essential), which may also be used for cooling the air in summer, when wanted (this may happen especially in inhabitable greenhouses).

Heat pumps can be made more efficient by "feeding" them with air pre-heated and pre-cooled by air-to-ground heat exchangers constituted by pipes embedded into the ground (prevalently, plastic, PVC pipes); as seen, at a depth variable, commonly, from 60 cm to about 2 m below the ground level. And the heat (or coolth) produced by the heat pump(s) can be distributed to the indoor space of the greenhouse not only by means of air channels, but also, as an alternative, through air channels embedded into the ground, so as to activate (call into play thermally) the grown soil. What is obtained, in the latter case, is a thermal system with a huge thermal mass (the ground), which is worth isolating with thermal insulation from the surrounding ground.

A further alternative is distributing heat or coolth in a radiative way. This can be obtained by distributing the heated or cooled air in a cavity beneath the pavement, the inertia of which depends on the quantity of mass of the floor system above the insulation. In a solar greenhouse, the same could be done with the massive back wall. In both cases, this strategy resembles that of the hypocaust system of the ancient Romans and could also be applied to cavities in the walls. The strategy is more suitable for low-temperature heating systems (of the kind of heat pumps and passive solar renewable energy systems) than to high-temperature ones (of the kind of boilers), but it is also compatible with high-temperature ones.

The most important fact about air systems is that air is the medium, and because air has a low thermal capacity, as already stated, it must be moved in large capacity systems of pipes or channels, in order to be effective for heat transport. This entails that the diameter of air pipes or channels has to be large (Dekay and Brown 2014; Grondzik et al. 2019), or that they have to be many.

As regards the air-system solution constituted by fans coupled with thermal electric resistors powered by photovoltaics, as seen, it is currently penalized by the less-than-excellent conversion rate of PV panels.

4.1.2. Water-based systems

Both the systems based on boilers and those based on heat pumps can distribute heat via a combination of radiation and convection. Water pipes, in this case, are the medium, but, because the thermal capacity of water, by volume, is high, the pipes of these systems have a small diameter and are easy to embed.

The most frequent combination for distributing heat and coolth in greenhouses are: (a) with regard to heating, ordinary radiators, that distribute heat and coolth mostly by convection (for a fraction of about 80%), and also transmit radiantly, although in a way that is too locally concentrated to be effective; (b) low-temperature radiators, which produce a higher ratio of radiant-to-convective transmission, and that, being distributed, are less problematic as regards the localized concentration of radiative heat, but that also, being more expensive, have a cost that is usually too high for bioclimatic greenhouses – except inhabitable ones; (c) high-mass or low-mass wall systems with pipes embedded into them. In the latter case, as in all air systems, the heat (or coolth) produced by the boilers (or heat pumps) can be distributed not only via air channels running through the indoor space, but also through air channels embedded into the ground. Indeed, horizontal radiant heating systems are usually the best choice when plants are involved, because vertical systems will primarily benefit humans (who tend to stand upright) rather than plants, and because humans are not likely to stand in greenhouses all night.

Systems of pipes embedded into the pavements for human usage can be low mass or medium mass. Those radiant systems can be obtained by placing insulation layers just a little beneath the embedded pipes and the pavements. Large-mass thermal systems can, instead, be embedded into the ground, at a level depending on how large the thermally controlled mass needs to be. In this case, insulating the ground below the greenhouse from the surrounding ground is likely a good decision, suitable for reducing heat losses to the external environment.

4.2. Heat recovery via air-to-air heat exchangers

An air heat exchanger typically receives the indoor air and the outside air, mixes their temperatures without mixing their mass, and outputs two separate air volumes of averaged temperature. In winter, typically, warm (and moist) air is taken from indoors, and cold (and drier) air is taken from outdoors; both volumes are driven by fans via pipes into the heat exchanger. After the heat exchange, cool air is expelled outside, and warmish air is given back indoors.

There are two types of heat exchanges: in the simplest and most common type, the moisture content of air is not modified, and in the other it is actively modified. In the latter case, during the winter, typically, warm and moist air is taken from indoors, cold and drier air is taken from outdoors, then cool and moist air is expelled outside, and warmish and dry air is released indoors.

Air-to-air heat exchangers can be used for heat recovery in actively heated greenhouses, where high air change rates are necessary, which would entail huge heat losses, and are commonly used in superinsulated buildings, in order to obviate the fact that, when superinsulation is in place, a great part of the building's heat losses will take place by air exfiltration and infiltration.

In crop-growing greenhouses, heat recovery of air is a high-end solution. Besides the case of high-end greenhouses, the adoption of air-to-air heat exchangers is, overall primarily attractive for inhabitable greenhouses, because it frees the inhabitants from the risks of stale air that may arise when the air changes towards the outside are maintained very low, for the sake of heat-loss reduction.

4.3. Passive and low-energy heating and cooling solutions based on the thermal exchange with the ground

Pipes embedded in the ground can be used in combination with a passive solar system, or, most often, with semi-passive solar systems where fans produce the movement of air; or they can be used in combination with active heating and/or cooling systems. How the pipes are laid out in the latter cases is similar. In the case of a purely passive system, the solution is different, because the air channels/pipes must be larger to limit the attrition of air as much as possible. The channels/pipes can be embedded into the greenhouse ground (see the GAHT systems – next section) and/or outside of it (as Canadian wells – section 4.4). Each solution has its own indications. In both systems, the most usual diameter of the embedded pipe is 12 cm.

On references

Heat exchange with the ground in greenhouses is a highly active research topic. Relevant references are Ghosal et al. (2004, 2005), Ghosal and Tiwari (2006), Tiwari et al. (2006), Ozgener and Ozgener (2010a, 2010b), Ozgener (2011) and Zhou et al. (2017). In Tiwari et al. (1997), the thermal behavior of a solar greenhouse coupled with an air-to-ground heat exchanger external to the greenhouse is studied analytically in quasi-steady-state mode. In Jain and Tiwari (2003), the optimal design of an air-to-ground heat exchanger for a greenhouse system is sought. In Peng et al. (2020), a ventilation system combined with air pipes embedded in the ground and with an attached greenhouse is presented.

4.3.1. GAHT systems

Air pipes can be embedded into the soil of the greenhouse, and/or in the soil below a building attached to the greenhouse, in order to create a fan-forced air loop of the type greenhouse-building-greenhouse passing through the ground, so as to improve the heat transmission to and from the ground itself. This loop activates a fan-forced seasonal use of the thermal mass that can be made more efficient by insulating the involved ground from the ground surrounding it. Plastic pipes evenly spread and evenly spaced into the ground are usually used for such a circuit. This strategy is named the ground air heat transfer (GAHT).

The pipes involved are usually PVC ones. In this case as well, they are most commonly embedded at a depth of about 1.2 m below ground. The pipes can be connected to form underground networks (see Figure 4.1A).

The solutions based on ventilation cavities in the ground rather than buried pipes need particularly reliable impermeable water and vapor barriers against the ground, in order to prevent the risk of indoor pollution by gas radon that is present there and that can be dangerous for the health, due to radioactivity.

The criterion for determining the set-temperatures of the fans charged with moving air, in the case of a semi-passive system or an active system, is positioning them a little above the target heating temperature and a little below the target cooling temperature. In the hypothesis that these two temperature levels are, for example, 2°C and 32°C, respectively, the set temperatures of the fan(s) may be, for example, 10°C and 24°C (Schiller and Plincke 2016).

There are four main possible arrangements of the pipes:

a) one loop of pipes set at a non-superficial depth – like -90 or -120 cm (see Figure 4.1A);

b) two loops of pipes at different depths in the soil – such as, for example, 60 cm and 120 cm (see Figure 4.1B). The advantage of this arrangement derives from the fact that embedding pipes at more than one depth improves the heat exchange with the ground;

c) a set of independent pipes. This solution can be adopted when it is simpler than the alternatives, at a constructive level (see Figure 4.1C);

d) the so-called barrel configuration, characterized by a main central collector from which the other pipes branch out (see Figure 4.1D). In this approach, a 55 gallon plastic drum is sometimes used as the main pipe.

In all cases, the connection between the pipes is made possible by T- or L-shaped adapters (see Figure 4.2).

In a GAHT system, the air loop can serve a passive solar system having the greenhouse as a solar connector – in which case, the overall system is of the passive type. Another possibility is that the embedded pipes are connected with a heat pump or another active heating system. In this case, the whole system is of the active type.

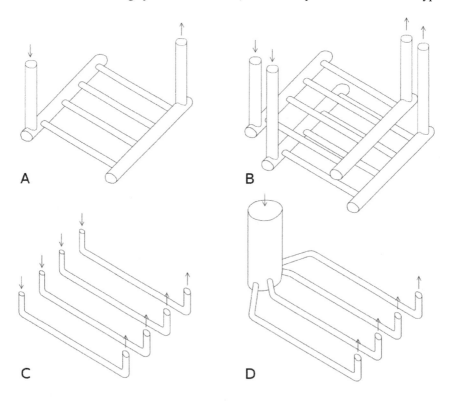

Figure 4.1. *Possible types of arrangements of the ground pipes. A. Two main collectors and a smaller-caliber grid of pipes between them. B. Two networks of pipes like in A, one above the other (commonly at a distance of the order of about 60 cm). C. Distributed system with no main collector. D. Distributed network connected to a barrel pipe*

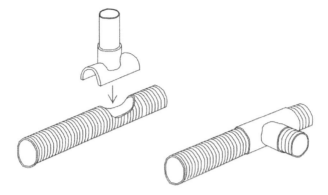

Figure 4.2. *Joining technique between plastic corrugated ground pipes*

On references

In Ozgener and Hepbasli (2005), the performance of a solar-assisted ground-source heat pump system for greenhouse heating is investigated. In Dai et al. (2015), the effect of a solar-assisted heat pump in combination with different heating operation modes is investigated. In Emmi et al. (2015), the effect of solar heat pumps in cold climates is analyzed.

4.3.2. *Ground-air heat exchangers – Canadian wells*

Another solution using ground-embedded pipes is based on Canadian wells. Canadian wells are constituted by ground-embedded pipes running outside the perimeter of a greenhouse and integrated into an open loop, going from outside in with no return – differently from what usually happens for grids of ground-embedded pipes exchanging heat with a heat pump (indeed, the latter exchange usually takes place in a closed loop). Typical lengths of the pipes in Canadian wells are between 12 m and 18 m for pipes of 12 cm in diameter, and the number of pipes used in a group is commonly 2 or 3. The pipes are placed in trenches, at a depth of usually about 1.2 m, for practical reasons of ease of excavation. At a thermal level, however, the deeper they are, the better it is.

After the pipes have been positioned in the trenches, the trenches are then refilled with soil.

Because the pipes in Canadian wells lie outside the greenhouse, the changes brought to the temperature of the soil into which they are embedded do not accumulate season after season and year after year, because what happens during the winter counterbalances what happens during the summer and vice-versa. In other words, the temperature of the soil, in this strategy, is not altered, in time, towards the most advantageous target, due to the presence of the greenhouse, but rather it is only altered (inevitably less substantially) by the outside air running through the pipes. The outside air has the effect of shifting the temperature of the soil around the pipes towards the mean seasonal air temperature at which the pipes are used, while the layers of soil further away tend to the mean yearly air temperature at the location in question. This is because the average yearly temperature of the air drawn into the pipes, in this case, is the same as the yearly ambient temperature. This entails that, more than in all other cases of ground-embedded pipes, in the case of Canadian wells, it is necessary to keep the pipes at a reasonable distance from each other – 60 cm should be considered an absolute minimum – to avoid mutually weakening their respective effect.

The air in the pipes can be drawn into the greenhouse by natural convection or by fans.

In winter, the system can be used to pre-heat the air entering the greenhouse (in winter, the temperature of the air in the pipes, despite being lower than the comfort threshold, will be higher than the environmental one), and in summer, it can be used to directly cool the incoming air. In this case, in temperate to cold-temperate climate, the temperature of the air coming from the pipes, to be usable, need not be modified.

When the stack effect is exploited for moving the air in the pipes, it is often necessary to design the air extraction by the chimney effect, which in turn requires taking into account a substantial amount of heat losses at the higher end of the stack. This is the reason why the stack effect strategy is, in the case of Canadian wells, more frequently adopted for cooling purposes than for pre-heating the incoming air.

When fans move the air in the pipes, air is frequently adopted as a medium both for cooling purposes and for pre-heating purposes.

The air in the pipes can be distributed into the greenhouse or circulated into an air-to-air heat exchanger to extract its heat (or coolth) while mixing it with the air taken from indoors. In the case of circulation by stack effect, only the option of air mixing is available. In the case of utilization of fan circulation, both options are available.

Especially when the stack effect is used for air circulation, the inlets of Canadian wells should be protected from thermal swings (so as to avoid the overheating of wells: in this case, they would produce a chimney effect in the direction opposite to the desired one – i.e. opposite to the one determined by the greenhouse), as well from the rain. Both things can be obtained by making sure that the pipes have access to clean and quiet air. Rain protection is usually provided by the cover at the end of each pipe.

An additional useful criterion is to avoid sloping the pipes towards the greenhouse. The pipes should, rather, be gently sloped in the opposite direction, away from the greenhouse. This avoids the water condensed in the pipes accumulating in the part of the pipes near the greenhouse. The described solution is suitable for combining with that of drilling holes in the lowest parts of the pipes and embedding those parts in gravel wells. In that manner the water condensed in the pipes can safely be drained into the ground.

4.3.3. *Considerations about the transfer of heat to remote masses by convection*

In greenhouses, the transfer of heat to remote masses by natural convection solves the difficulty of increasing the masses when there is no possibility of increasing the solar aperture; however, to be put to work, this transfer strategy requires solving issues like maintainability, healthiness and pressure loss by friction along the air flow path.

The rationale of the strategy is based on the stack effect.

Where there are temperature differences between zones and between openings, and where there are differences in height between openings, the air can be moved from the greenhouse to the building (see Volume 1, Figure 1.33-left), or from the frontal zone of the greenhouse to the back zone of the greenhouse (see Volume 1, Figures 1.33-centre and 1.33-right), depending on the situations. The air movement takes place both if the greenhouse (or, in the latter case, the frontal zone of the greenhouse) is warmer than the building (or, in the latter case, the back of the greenhouse) and if the greenhouse (or, in the latter case, the frontal zone of the greenhouse) is cooler than the building (or, in the latter case, the back zone of the greenhouse). Seen from the perspective of the greenhouse, if the greenhouse is warmer than the building, the warm air pushes upwards from the greenhouse to the building (or, in the latter case, to the rear zone of the greenhouse) and if the greenhouse is cooler than the building, the cool air pushes downwards from the greenhouse to the building (or, in the latter case, to the frontal zone of the greenhouse). But, because the transfer of heat to the building mass is wanted only in situations of heating need, and the transfer of coolth to the building mass is wanted only in situations of cooling need, it is usually better to prevent the air flow determined by the bottom-up air movement from the greenhouse (or, in the latter case, the frontal zone of the greenhouse) during the summer, taking place mostly in the daytime, as well as the air flow determined by the top-down air movement from the greenhouse (or, in the latter case, the frontal zone of the greenhouse) during the winter, taking place mostly at night.

4.3.3.1. *Horizontal rock-beds, horizontal block channels and ground-embedded air pipes*

Channeling the air through horizontal rock-beds is a proven solution for convective heat transfer (see Volume 1, Figure 1.33). The main components of the system are as follows: (a) two cavities across all of the greenhouse depth (see Volume 1, Figure 1.33, center and right), or both along the greenhouse depth and along the building depth (see Volume 1, Figure 1.33, left); (b) horizontal grilles or open-jointed planks for dividing those cavities from the spaces above; (c) two walls with holes to contain the rock-beds; and (d) the rock-beds, possibly with channels included, to reduce the friction of air; or, as an alternative, hollow concrete blocks laid out to form channels between the two cavities (see Volume 1, Figure 1.34). The rock-beds may be insulated beneath, in order to reduce the thermal losses towards the ground, while profiting from the thermal inertia of the ground itself; or they may be insulated beyond and beneath, so as to make, additionally, the heat transmission take place mostly by convection (this is advantageous when more heat is wanted elsewhere compared to at the ground floor – for example, at the first floor of an attached building), or they may be insulated only above (to couple the previous situation with the fact the temperatures are maintained more stable through the year, as a result of the relation with the ground). Indeed, when the rock-beds (or the mass of hollow concrete block) are uninsulated below, they – due to the heat exchange with the ground – constitute a hybrid storage system partially working on the basis of a daily time period, and partially working on the basis of a seasonal time period. The problem of the formation of odors due to bacteria, deriving from the fact that the air in the rock-beds is the same as that in the room, is well known, but in this scheme, there is little possibility of cleaning the ducts in the beds without disruptive interventions. Therefore, the technique in question does not appear to be very suitable for updating today's residential standards – at least, without introducing substantial changes. This issue pertains to attached greenhouses, but it is less important in stand-alone greenhouses not aimed to principally host human beings. In any case, updating the systems to the most modern performance criteria would require making the cavies in the ground accessible for periodical cleaning, but this is disruptive with respect to the original construction conception.

The situation is different as regards the solution using the concrete hollow block channels, because designing such a system for maintenance and accessibility is likely to be easier. Also, the quality of this solution can be more easily controlled and produce more predictable performances than the rock-beds. In it, the hollow blocks are laid out on their long sides, in one, two or three layers, so as to make the cavities continuous. The main difference between this solution and the rock-bed solution lie in the greater constructability and accessibility of it. The utilization of the system in question is especially championed by Kachadorian (1997).

Horizontal rock-beds (or channels in hollow blocks) can constitute a stand-alone heat storage system or be coupled with a vertical rock bed heat storage system (see Volume 1, Figure 1.33, center). Also, vertical-only storage systems are possible (see section 4.3.3.4.).

On references

Studies based on the use of rock-bed thermal masses coupled with passive solar greenhouses are Asa'd et al. (2019), Gourdo et al. (2019) and Bazgaoua et al. (2020). Kürklü et al. (2003) investigate the coupling of a rock-bed with a polyethylene greenhouse.

4.3.3.2. Horizontal cavities

An alternative to rock-beds and horizontal block channels is constituted by accessible channels built with concrete walls or masonry walls oriented in the transversal direction of the building (the direction of the air flow), possibly arranged so as to allow accessibility and maintainability of the cavities. To make this possible, cavities with a minimum height of 80 cm are suitable. From a thermal point of view, this kind of solution exploits the space less efficiently than that of hollow block channels, because the mass, in this case, is less concentrated in space; however, the possibility of making these cavities more accessible and maintainable to avoid the microbes and fungi issue more easily outweighs the disadvantages.

To exploit the available space, it is highly beneficial that the bottom of the cavity is also thermally exploited with a thick layer of concrete. This thick concrete layer can be part of the support of the underground walls (in which case it needs to be reinforced), or it can be cast between them, in order to make the walls supported directly on the ground via the first screed (in which case the screed need not be reinforced, but may be reinforced mildly).

4.3.3.3. Vertical rock-beds and hollow-block channels

In experiences from the 1970s and 1980s, vertical heat storage rock-beds constituted an alternative to horizontal rock-beds heat storage systems, or were coupled with them, with which they share the same problems of growth of microbes and fungi, uncleanability and unmaintainability. The system in question is composed of (a) a full-width cavity, usually at the side of the vertical rock-bed, but possibly below it, and a full-width cavity on top of it; (b) grills or planks letting the air flow through, in a position (horizontal or vertical) appropriate to the position of the cavity and (c) the rock-bed, possibly including channels obtained by suitably positioning the rocks. As seen, when there is a continuity between the horizontal rock-bed and the vertical one, the possibilities are (1) having three full-width distribution cavities, the central of which is not in communication with the spaces above, but only with the two rock-bed channel systems, (2) having only two full-width cavities at the extreme ends of the systems and (3) making the rock-bed channels continuous.

The solution of vertical block channels is analogous to the solution of horizontal ones, from the point of view of the components, and as regards the function, it is analogous to that of the vertical rock-beds. This solution is more health-compatible, and its results are more reliable than the rock-bed solution, but it is almost as non-visitable and non-maintainable as that. It is, however, simpler to build, for the reason that no full-width distribution cavity is needed at the bottom of the system. This in turn is due to the fact that its distribution function can be performed by a "jigsaw" arrangement of blocks at the wall base (i.e. an arrangement with blocks and voids alternated).

4.3.3.4. *Vertical cavities*

The vertical cavities should be visitable, as should the horizontal ones, but, in this case, they are easier to attain than with rock-beds and hollow-block channels. However, vertical channels are less likely to be overall advantageous than horizontal ones, because they are likely to occupy a more useful and costly space. In addition, the hollow vertical channels may be accessed with difficulty unless some panels are devised for dedicated access; but this, ultimately would reduce the surface available for thermal mass. However, when both horizontal pipes/channels and vertical pipes/channels are present, the function of storing most of the heat may be given to the horizontal channels, while the function of increasing the air pressure head for creating the stack effect – and therefore the movement of air within the system – may be given to the vertical pipes/channels. In this case, a few channels may suffice to produce the needed air flow capacity, which would involve saving space in the zones above floor level, or the channels may even be left with no mass around, in order to retain only the function of air "movers". In the latter solution, the coupling of horizontal channels with vertical ones can be very advantageous (see Volume 1, Figure 1.33, center).

4.3.4. *Surface air-to-ground heat exchange (experimental)*

It is possible to use the surface ground around the greenhouse as an air-to-ground heat exchanger: this can be done by driving the air flow by stack effect into an open-loop from the "exchanger" to the top of the greenhouse (see Figure 4.3). Because the air loop is open, the strategy is useful in the hot seasons and the intermediate ones, as a cooling strategy. The strategy requires a trench to be dug at a façade of the greenhouse (usually, the front façade) and that the air flow is compelled to pass through the trench before entering the greenhouse. The trench may be covered by grids to make it walkable.

Heating and Cooling Systems; Watering Systems 203

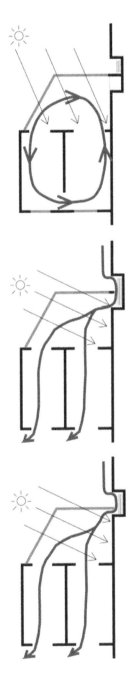

Figure 4.3. A trench, possibly containing water, can be used as a summer ground-to-air heat exchanger (left and center). In winter, the trench may be emptied dry, or only the internal part of it may be emptied dry

Construction-wise, there are two ways for devising this system. One is building the trench (with or without retaining walls, depending on the economic investment) entirely outside the greenhouse, separated from the foundation wall by a layer of insulation. In this case, the simplest way for creating the entrance for the air of the open-air loop is to cantilever the greenhouse façade out of the foundation wall and providing the bottom frontal areas of the greenhouse with operable openings that remain out of the way of the transparent front (see Figure 4.3, left). The other solution is creating the trench partially inside the greenhouse and partially outside (see Figure 4.3, center), joining the two parts via openings in the foundation wall, keeping the foundation wall insulated, and making it closeable in winter with operable insulated openings.

In both cases, the trench can be filled with water, in order to add the possibility of evaporative cooling to the system. The water may then be drained away in winter, when the openings of the system are closed. This would be optional in the former case, but would likely be almost unavoidable in the latter, which otherwise would cause strong thermal losses in the outer part of the trench (see Figure 4.3, right).

A more complex solution is constituted by creating a fixed insulated partition between the two parts of the trench hosting the water, so as to create the possibility of making each of them drainable out autonomously. This may be useful in keeping only the indoors filled in winter, in order to avoid the humidity of the latter. As an alternative, when both parts are kept filled with water, the solution will be useful to allow for the possibility of keeping the two water parts at different temperatures and prevent heat losses from the system via the outer part of the pool.

4.4. Auxiliary heating systems

Auxiliary heating is not always needed in bioclimatic greenhouses. One reason for this is that in winter, in many cases, the temperatures can be kept at levels that are acceptable for both plants and human beings, even when no auxiliary heating plant is used. Another reason is that when the full winter arrives, crop-growing activities may be stopped, and low temperatures may be accepted for plant survival.

But in some cases, heating systems are indeed advantageous. Often, they are the only solution allowing for year-round growth in some conditions, and/or for year-round human thermal comfort.

The share of heating that is by far the most difficult to provide is the last one remaining after all possible solar measures have been pursued. Most greenhouses in cold or temperate climates renounce to control temperatures in the coldest period of the winter – giving up on providing shelter for plants – or to adopt an auxiliary heating system, meant as a backup system suitable for assistance in the possibly long and cold spells of winter days with cloudy skies.

A super-simplified way for calculating the heating loads of a greenhouse (i.e. the amount of energy required by it) is the following:

Required heat [KJ] = (Wall_area + roof_area) x (T_inside – T_outside) x 1.1.

In general terms, the energy consumption depends on the thermal loads (which in turn depend on the greenhouse; more specifically, on their solar gains and thermal losses) and on the efficiency of the heating system, which largely depends on its type and sizing.

For greenhouses, auxiliary heating systems can produce heat by combustion – like boilers do – or by more energy-efficient strategies, like heat pumps linked to ground-to-air heat exchangers and driven by electricity produced by photovoltaic panels. Boilers can produce higher temperatures than heat pumps, and this influences how heat should be distributed in the two cases. But heat pumps are more efficient, and can also produce coolth, which can be distributed in the same way that heat is distributed.

The two principal ways that heat and coolth can be distributed are through air (by convection) and surfaces (by radiation and convection conjointly). When heat is distributed radiantly, the system could be of the high-temperature type or of the low-temperature type, and can have high thermal inertia or low thermal inertia.

The best combination in each specific case depends very much on the objectives of the greenhouse. But in any case, it should be considered that the return on investments of the heating plants for solar greenhouses is slower than for buildings, or even than for conventional, non-solar greenhouses. This happens because the heating plants in a solar greenhouse tend to be used much less than in a house or in a conventional greenhouse, and this makes it more difficult to make them pay for themselves in time.

The implications of the main types of auxiliary heating systems are presented in the following sections.

On references

Some books belonging to the literature about passive solar systems and concerning the self-design, self-construction and usage of passive solar systems are Szokolay (1975), Scudo (1993 – in Italian) and Smith (2011). Reviews of heating technologies for greenhouses are Sethi et al. (2013) and Cuce et al. (2016). In Ahamed et al. (2018), a computational model simulating the auxiliary heating requirements of film-enclosed solar greenhouses is presented. In Lazaar et al. (2015), a comparative experimental assessment of solar and conventional heating systems is presented. Shan et al. (2016) and Hematian et al. (2019) analyze strategies for the usage of heat pumps in greenhouses. Lu et al. (2017) focus on floor radiation in greenhouses, and Bouadila et al. (2014) focus on packed-bed solar air heaters. In Liu et al. (2018), the use of a sunspace-based hybrid heating system for reducing energy power in dwellings is analyzed.

4.4.1. Electric heating

Electricity is the most convenient form of energy at the level of transfer and management, and electric heating systems are low-cost, but electric heat is not efficient when the electricity has not been produced efficiently. What is inefficient is not the conversion from electricity to heat in itself, which, on the contrary, is very (almost 100%) efficient, but the production of electricity via combustion engines (which has an efficiency well below 50%). However, combustion engines are still today the most common source of energy when the electricity comes from the grids. Another inefficient solution is to couple photovoltaics directly with electric resistors to generate heat, because this strategy, due to the level of efficiency of photovoltaic conversion, is much less (around three times) efficient than solar passive water heating (based on the use, for example, of solar water heating panels) (Matuska and Sourek 2017).

For these reasons, electric heaters are considered especially suitable for small greenhouses.

Electric heaters are usually coupled with fans into small combo units that can be carried around a greenhouse. In this case, it is the air that gets heated. Therefore, the lack of efficiency of the system will be further increased if the greenhouse is leaky, that is, not sufficiently airtight.

Air-based electric heating is also unsuitable for producing substantial localized heat. This is also due to the fact that the air jet exiting a combo unit cannot be directed towards the foliage: this would distress the plants and burn their leaves.

Another common (low-cost) configuration for electric heaters is constituted by the so-called unit heaters – that is, combo units combining the heating system and a fan. When a unit heater has to be coupled to a perforated plastic pipe (usually, of polyethylene, for greater transparency and low cost) (see Figure 4.4), the fan should be strong enough to move the air despite the high friction, and should be more appropriately defined as an air blower. It should be noted that unit heaters can also be coupled to heating devices different from electric heaters, like boilers connected by hot-water pipes.

Unit heaters can usually serve greenhouse lengths up to about 18 m. In these configurations, a unit heater blows air into the perforated pipe, that (a) can be placed high in the greenhouse, when the blower can be used both for heating (which happens when the heating unit is on and the vent near the blower is shut – see Figure 4.4, left) and for cooling in winter (which happens when the heating unit is off and the vent near the blower is open – see Figure 4.4, center – so as to draw in the outside air) or (b) it can be placed low in the greenhouse, when it is only used for heating. This is because the air cooler than that in the greenhouse is more efficiently distributed to the greenhouse from the top, because it tends to sink; while the air warmer than that in the greenhouse is more efficiently distributed from the bottom, because it tends to rise. When both goals have to be pursued (at different times, of course), the top position is more practical, because the need for cooling by drawing air into the greenhouse can occur in both summer and winter, while the need for heating by circulating the air traversing the unit heater can occur only in winter.

In this configuration (see Figure 4.4, center), air cooling in winter can be pursued by opening a vent in the vicinity of the blower, in order to draw cool air from it into the greenhouse, as well as keeping some outlets in the greenhouse opened, in order to favor the exhaustion of air derived from the positive pressure in the greenhouse. When air cooling is necessary in summer (see Figure 4.4, left), the air circulation in the greenhouse must instead be stronger, and therefore, it is necessary that (a) a fan draws the air at the end of the perforated pipe, rather than pumping it in and (b) that the fan is not detached from the vent, but it is attached on the greenhouse envelope. Also, in these cases, some of the inlets in the greenhouse must be opened to favor the expulsion of air.

The unit heater for blowing air into the pipe and the fanned vent for drawing air from the pipe can be combined at the two ends of the perforated pipe for greater seasonal flexibility.

In single-bay greenhouses, systems of one blower and one perforated pipe are enough, while in multi-span greenhouses, one unit heater and one perforated pipe for each bay are needed.

Another (low-cost) strategy – not entailing perforated plastic pipes – for distributing heat convectively in combination with unit heaters is that of combining the fans of the unit heaters with fans for creating a horizontal air flow (HAF systems).

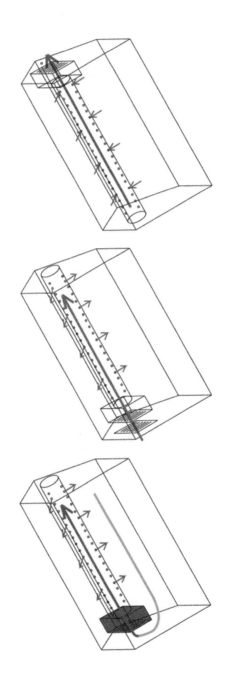

Figure 4.4. Example of perforated fan-tube systems. Left: unit heater blowing air into a perforated pipe in a closed loop, for winter heating. Center: fan drawing air from the outside, through a vent, and blowing air into a perforated pipe, in an open loop, for winter cooling. Right: fan drawing air from a perforated pipe and expelling it through a vent, in an open loop, for summer cooling

An electric solution suitable for producing high localized radiant temperatures stably is that of the portable ceramic disk heater, which involves the use of more mass, and is of the radiant type.

4.4.2. Common stoves

Wood stoves and pellet stoves are simple and efficient in their modern versions, but require constant feeding and tend to produce too high localized temperatures for the plants. For that reason, stoves should be kept reasonably away from the plants, as well as from parts of the greenhouse – like plastic panel enclosures – that can be damaged by their heat. Another drawback of stoves is that they require flue pipes for exhausting the gases after combustion.

4.4.3. Rocket mass stoves

A rocket mass stove is a very interesting option, because it entails heating a mass from the inside in a short rush of hours, then letting the mass float thermally during the night and release the heat slowly. It is a strategy allowing stable low-temperature radiant heating without having to feed the stove continually. However, in its various versions, the rocket mass stove is a complex solution at a constructive level. A description of the system is beyond the scope of this book, but there are easily available sources describing its rationale. The traditional variant of the system, inspired by the Russian stove, has been described in Kern (1975). The modern version of the rocket-mass stove, reaching much higher temperatures and greater efficiency, thanks to a high-temperature combustion loop, has been described in Schiller and Plincke (2016).

On references

In the book by Kern (1975), among other things, the rationale of the massive stove construction is presented.

4.4.4. Water systems coupled with burners or heat pumps

4.4.4.1. Fan coils

A fan coil is a water-to-air system that transfers heat from the water to the air through a water-to-air heat exchanger combined with a fan. It is not based on movable devices, except the fan, but it allows the water to be heated through strategies that are more efficient than electric thermal resistance, like combustion (of wood, pellets, a gas, or liquid fuel) or based on the use of heat pumps, possibly driven by photovoltaic electricity. However, all these strategies are substantially more complex than those based on electric heaters.

On references

In Matuska and Sourek (2017), the performance of a water heating system coupled with PV panels and heat pumps is analyzed.

4.4.4.2. Radiant systems

To provide low-temperature, efficient radiant heating, a water system can also be combined with pipes embedded in the ground, or in some areas of the floors or pavements, or in the walls. In those cases, the system, especially in greenhouses, is usually high mass and not quickly responsive to the evolution of the temperatures signaled by the thermostats. Its action, usually, should therefore be planned, possibly also considering the weather forecasts.

But water pipes can also be used as low-mass systems, for greater responsiveness and flexibility. One solution for obtaining this is suspending the water pipes overhead. This is the most common active radiant heating solution in greenhouses. Another solution is embedding them into low-mass construction elements, like, for example, light partitions, or concrete screeds laid on top of insulated supporting floors. An example of this solution is constituted by a screed of 4 or 5 cm of sand embedding the pipes and sitting above an insulation layer and a concrete slab.

Another type of a low-mass system is constituted by earthen water pipes laid out on the ground, which, however, are likely to be at serious risk of damage in work environments involving plant growing.

4.4.4.3. Gas heaters

Natural gas or propane heaters have the advantage of low cost fuel and efficiency, but the disadvantages of polluting the environment and complexity. The complexity derives from requiring lines and flue pipes – a chimney or a vent pipe. This is because carbon dioxide (used by the plants) must be removed from the indoors to reduce the risk of retention of pollutants (including traces of carbon monoxide).

4.4.5. Active systems using renewable energy sources

Another route is using heating plants exploiting renewable energy sources. The most common case is that of solar water panels. The panels can be active (the water, in this case, is circulated by a pump) or passive (integrating the water tank and letting water circulate passively), and they can be of the flat-plate type or the evacuated type. The panels can be situated out of the greenhouse, or within the greenhouse, at the cost of a certain reduction of the solar radiation in entry operated by the transparent panels (Bargach et al. 1999). When the panels are kept outside, it is essential that they do not reduce the solar aperture of the greenhouse, because this would happen at the expense of the overall thermal efficiency of the greenhouse in winter, and in a greenhouse context, this may not be as easy as it sounds when the surface area of the panels is large.

On references

In Bargach et al. (1999), the performance of a flat-plate collector in the context of a greenhouse is analyzed in detail.

4.4.6. Heat pumps

Another efficient option for contributing heat to a greenhouse actively is the use of heat pumps. This is because heat pumps are very efficient – their efficiency, calculated on their need for electricity, is usually between about 3 and about 5, and it needs a medium – a thermal well – as much as possible stable, and protected. The position of the heat exchanger is sometimes in the ground, sometimes in trenches and sometimes in the open air (these options have been mentioned in decreasing order of efficiency). The electricity required by the system can be provided by photovoltaic panels, which, ideally, should be positioned so as not to cast shadows on the greenhouse during the winter, in order to avoid reducing the solar aperture (Ozgenera and Hepbaslib 2005).

On references

Ozgenera and Hepbaslib (2005) investigate the performance of a solar-assisted heat pump looped with the ground.

4.5. Auxiliary cooling systems

The lowest-energy semi-active cooling strategy is constituted by the simple, untreated convective exchange with the ground through buried pipes, which can also be used during the cold season for heating. The medium in the underground pipes can be constituted by water or air. When air pipes are used for cooling the soil during summer, it is common that the water vapor condenses on them, and to avoid clogging, the pipes, as seen, should usually be sloped, and possibly drained through holes in the zone near the elbow at the lower end of the slope – for example, in gravel beds located there.

The air in the pipes is most often moved by fans, and how much the coolth transmitted via the pipes is or is not suitable for cooling a greenhouse without further thermal treatment depends on the temperature of the ground. In cold and temperate climates, the temperature of the ground at a depth lower than 2 or 3 m is usually low enough to cool a greenhouse directly: but in hot climates, this temperature is usually not low enough to do that. In those cases, the air in the pipes may be used to pre-cool the air to be served to a chiller (a heat pump working in reverse mode) for cooling.

The lowest-cost and lowest-energy consumption active cooling strategy for greenhouses is constituted by evaporative blowers based on fans powered by electricity blowing air through wet pads into the greenhouse space. The action of the fans can push the air through the pads, or draw the air from the pads. In the first case, the fans are located near the pads, externally to them, and the pressure in the greenhouse is positive, and in the other case, the fans are located at the opposite side of the greenhouse with respect to the pads, and the greenhouse is in negative pressure. Evaporative cooling pads are especially indicated in hot–dry climates: in these cases, they can determine a temperature reduction of the order of 10–15°C. Of course, in any case, their action increases the relative humidity of a greenhouse, because the evaporation process takes place in constant enthalpy. Therefore, in these cases, it is necessary that a share of the moist air is expelled from the greenhouse, in order to keep the indoor relative humidity level in control.

The constant-enthalpy nature of the process does not make this strategy suitable for the most humid climates.

The lowest-cost cooling system is constituted by simple ventilation with air fans, and has been commented in Volume 1, section 1.6.5.4.1. The use of fans may be necessary for commercial greenhouses when the climate conditions are too hot for passive-only ventilation. But in a well-designed greenhouse, the use of perforated polyethylene pipes in combination with fans is likely to be necessary only in cases that are very difficult to address. In temperate climates, the use of perforated pipes is commonly adopted by combining several uses of the pipes, like winter ventilation and summer ventilation, as well as, possibly, convective distribution for winter heating (see section 4.4.1).

A high-energy consumption strategy for cooling greenhouses is constituted, as anticipated, by air conditioning, that is, utilization of chillers coupled with air-to-air heat exchangers. This strategy is wasteful of energy because there are many air changes of greenhouses, which is incompatible with air conditioning. A more efficient use of chillers for cooling entails combining them with radiant heat transmission, which vertical radiators or unearthed pipes may perform. Due to the high level of humidity of greenhouses, however, radiant cooling causes condensation of water vapor, and this should be taken into account in the detailing of the system. For example, when pavements are cooled via embedded water pipes, this should be taken into account in choosing the finishes.

On references

Cooling pads constitute a system of major importance in hybridly cooled greenhouses. Ganguly and Ghosh (2007) model and assess a fan–pad ventilated system for greenhouses. Aljubury and Ridha. (2017) analyze the coupling of an evaporative cooling system with geothermal energy. McCartney et al. (2018) present an innovative misting system for greenhouses in the context of hybrid ventilation. In Jain and Tiwari (2002), an optimal design of pad-based evaporative cooling systems is sought.

4.6. Integration of photovoltaic panels in greenhouses

Integrating photovoltaic panels in greenhouses is difficult if no opaque surfaces for integrating the panels – like, for example, a portion of opaque roof – are available. This is because when opaque photovoltaic panels are placed outdoors, beyond the transparent panels of greenhouses, they make shadows on them, without substantially reducing their thermal losses. This decreases the thermal efficiency of greenhouses themselves. And if the photovoltaic cells are integrated into the transparent panels of the greenhouses, the solar gains they produce besides electricity make the transparent enclosures warmer, increasing the thermal losses of the transparent panels towards the outside. This is much less of a problem in the case that the transparent panels are tiltable like fins, and used for summer solar shading, because their adjustability allows them to optimize solar shading and electricity production during the summer.

Another thing to consider is that when opaque photovoltaic panels are placed outside an opaque envelope, like, for example, the opaque part of a roof, they should be kept detached from it, so as to receive ventilation at the back, in order to avoid getting too hot during summer, which could decrease their efficiency up to stopping them from working altogether.

Placing the photovoltaic panels inside greenhouses does not entail thermal drawbacks, except the reduction of impinging solar energy due to the external glazing. In the case of growing greenhouses, the practical drawback is that the panels will make shadow on the plants anyway, and in the case of inhabitable greenhouses, they will subtract precious living space.

Despite these difficulties, due to the flexibility of electricity and the high demand for it, the integration of photovoltaics into greenhouses is presently very common, and used in combination with a large variety of demands, from fans to heat pumps and all sorts of control actuators.

On references

Corkish and Prasad (2006) review photovoltaic systems for buildings. Zhang et al. (2020) investigate the sizing of photovoltaic panels for integration in a solar greenhouse. Cossu et al. (2014) assess the solar radiation distribution in a greenhouse with photovoltaic roof, and its effect on crop growth, and Cossu et al. (2018) assess, more specifically, the solar radiation distribution in commercial photovoltaic greenhouse types. In Alinejad et al. (2020), the effects of an adjustable photovoltaic blind system for greenhouses are investigated. In Bambara and Athienitis (2019), the performance of a semi-transparent photovoltaic greenhouse enclosure is analyzed. Aste (2008) (in Italian) is a solid reference on photovoltaics in architecture.

4.7. Integration of passive solar heating panels in greenhouses

The most common objective for which passive solar heating panels may be integrated into greenhouses is that of warming water. In theory, passive solar heating panels might also be employed for warming air; however, this is something that a solar greenhouse usually can do well by itself. Additional heated air is beneficial for a greenhouse only if it is generated by a solar air system that does not reduce the solar aperture on the greenhouse – in other words, that is placed anywhere other than on (in front of or on top of) the greenhouse: for example, at its feet – a stalwart of the literature for passive solar systems of the 1970s, which also makes possible to distribute heat from the solar collector to the greenhouse by thermosyphoning, and to increase the solar gains of the greenhouse without increasing its thermal losses (Mazria 1979; Winter Associates 1983).

Indeed, at least in theory, the major advantage of coupling solar heating panels with a greenhouse is that they can indirectly extend the solar aperture of the greenhouse at will, because they can be positioned so as not to obstruct the solar "view" of the greenhouse itself – that is, away from the glazing. However, in practice, cost considerations related to heat transfer often play a major role in discouraging such a strategy.

The integration of water solar heating panels poses challenges similar to those posed by opaque photovoltaic panels: that is, they make shadows on the glazing and lay out the basis for worsening the thermal balance of a greenhouse even when they are well exposed. Therefore, their most appropriate use is at the outside face of opaque enclosures, such as opaque roofs; however, unlike photovoltaic panels, they can be integrated into the envelope of the greenhouse (so as to open the possibility of simplifying the construction solution), because their efficiency does not decrease when they get hot.

On references

Mazria (1979) is one of the most successful and influential books on bioclimatic building design up to this day. The book by Shurcliff (1979) constitutes a lively and mind-opening presentation of innovative ideas for low-cost solar heating from the 1970s, still of interest today. The book by Winter Associates (1983) constitutes a professional-oriented handbook presenting the basics of passive solar design, construction information.

4.8. Watering systems

Except for the periods of hot weather requiring full ventilation and maximization of the openness of a greenhouse towards the outside, plants in a greenhouse require much less watering than plants in an open field, due to the fact that a greenhouse is a closed system trapping in the water vapor produced by the evapotranspiration of the plants and the soil.

During the hot season, greenhouses usually are largely kept open, and, in those conditions, the plants need more watering. In this period, it is normal for the plants to require watering each day; while in the heating season, they may need watering once or twice a week.

Watering requirements depend also, and substantially, on plant species. Different plant species have different environmental (thermal, lighting, moisture-related) requirements regarding the watering schedules.

When the plants are in pots, it is easier to make sure that the soil receives the correct amount of water. A robust rule of thumb for doing that is watering the soil until the absorption response keeps being intense (provided that the soil mix is correct – i.e. provided that it is not too absorptive or too little absorptive), which can be obtained by watering each pot until water appears in the dish below it. This criterion should be combined with the fact that the growing soil should be watered when it becomes dry: this helps to fortify the plant roots.

On references

Criteria for quantifying the watering requirements of crops are presented in Allen et al. (1998). Reference to watering systems is especially abundant in the books focusing on the crop-growing aspects of greenhouses, like Pierce (1981–1992), Smith (1992), Taylor (1998) and Beytes (2003). The interaction between watering strategies and greenhouse microclimates is investigated by Li et al. (2020). The studies by Zaragoza et al. (2007) and Weichun et al. (2008) focus on water-saving strategies for greenhouses.

4.8.1. *Most common water sources*

4.8.1.1. *Tap water*

Tap water has the disadvantage that it is chlorinated, which is negative for the health of the plants. But fortunately, chlorine evaporates easily from the water, if the water is not closed with a lid – especially, an airtight lid. Therefore, the strategy for treating tap water is sitting it in containers – pools, drums – for some hours – most commonly, one night – before using it. If the water containers are not placed in the open, but are located in the greenhouse, there is the additional advantage that water will be given to the plants at the temperature of the indoor environment, which is the best thing for the health of the plants themselves, especially in winter. The disadvantage of the strategy of letting chlorine evaporate from the water is that it takes plenty of containers to host all the water needed for a watering session. On the on the other hand, tap water entails the huge advantage that, because it is pressurized, the containers can be positioned at a certain height to allow water distribution by gravity without the need for any additional energy input by pumps.

4.8.1.2. *Rainwater*

Rainwater has the advantage of purity over tap water, but it has to be harvested when there is a possibility, which does not necessarily guarantee that it suffices for covering the needs at all seasons. In places with dry hot seasons, in particular, to make the harvested water suffice for all seasons, including the hot ones, water may have to be collected in rainy periods to be stored in wait of the hot, dry season, and this may require huge containers and collection surfaces subtantially greater than the area occupied by the greenhouse.

The most important data to be considered for calculating the water collection potential is the average yearly rainfall, the average rainfall for each month, the surface available for collecting water and the efficiency of collection, which depends on the tendency to absorption of the collection surfaces and on the speed of collection, which depends on the roughness (it is hindered by it) of the surfaces involved. A barebone estimative approach can be based on the hypothesis that the collection efficiency will be of the order of 50%, which is a pessimistic stance for a greenhouse.

The ideal surfaces for rainwater collection are the roofs, because they are usually the surfaces that are cleanest from dust. Water-impervious pavements in open spaces can be used for collection as well, but they are likely to collect lots of dust and dirt during the first phase of the runoff. This can be addressed with downpipes completed with an inverted Y-shaped internal separation chamber suitable for separating the first runoff water from the next flow (Kinkade-Levario 2007), at the expense of greater cost and complexity.

On references

Kinkade-Levario (2007) is a good reference about rainwater collection, suitable for practitioners.

4.8.2. Water containers

Containers can be of many types, but the most common ones are pools or drums. The advantage of the drums is the possibility of placing them at certain height levels, in order to exploit the pressure created by gravity and assure energy-free distribution. The advantage of the pools is the ease of dealing with large water volumes, and their main disadvantage is the space they occupy. An interesting way for producing durable low-cost containers at a low cost is to use ferro-cement (see, for example, UNHCR (2006) and Ludwig (2009)).

On references

A good book about rainwater storage, suitable for practitioners, is Ludwig (2009). The construction of ferro-cement tanks is more specifically dealt with in UNHCR (2006).

4.8.3. Water distribution

The technical strategies for water distribution are numerous. The ones which will be dealt with in the following sections are the most easily manageable among them.

On references

Hanan (1998) details watering strategies.

4.8.3.1. Hose

Distributing water with a hose requires the water to be under pressure, and there to be lots of frequent human labor, but it is the cheapest solution besides watering with buckets.

4.8.3.2. *Drip system*

In the drip system, water is distributed by means of pipes that are bored with small holes at regular intervals, and (usually) are laid out at the surface of the soil in a distributed fashion, in order to allow the water to be released uniformly across the surface of the soil in the growing beds. The dripping pipes can be man-made by creating small holes along the sides of a small tube or factory-made.

To distribute the perforated pipes evenly, a net is formed, constituted by primary pipes and secondary, smaller pipes, having a diameter of the order of 1–2 cm. The diameter and number of the primary pipes depend on the diameter and number of the secondary ones. One advantage of the drip system is that it is very parsimonious in terms of water usage, because it does not entail wetting the plant leaves. However, while some types of plants can profit from the fact that their leaves are not wet, some others do not.

Not watering the leaves brings the added consequence of lowering the relative humidity in the greenhouse with respect to what it would be with the other watering methods. But for this strategy to work well, it is necessary that the outlet of the water container is provided with a valve preventing backflow (which may bring back at the source the organic and inorganic substances in the soil, at the risk of polluting the water in the containers).

Another drawback of the drip system is that it requires maintenance to keep the pipes, and the holes in the pipes, unclogged by the soil, where they are embedded.

The water distribution can be performed manually or with timers.

Commonly, drip irrigation systems are made with plastic outlet pipes (PVC or black MDPE, derived from polyethylene) between 2.5 and 4 cm in diameter as primary pipes, combined with water tanks of a capacity between 200 and 1,000 liters, mounted on platforms 1.5–2 m high. Ideally, pipe lengths of 17 m should not be exceeded.

Ghosh (2020) reports the following simple formula for calculating the estimated peak water requirement for greenhouse irrigation:

$$WR = [A * B * C * D * E] + C_{wr},$$

in lpd/plant (i.e. liters per day per plant), where:

A is the pan evaporation inside the greenhouse (in mm/day);

B is the pan factor (0.7);

C is the cropping area/plant (m^2);

D is the crop factor (1 for full-grown crops);

E is the wetted area (0.3 if widely spaced, 0.7 if closed spaced);

C_{wr} is the crop water requirement (lpd/plant).

On references

A clear how-to reference on drip irrigation is Müller et al. (2015b).

4.8.3.3. *Overhead sprinkling*

An alternative to the drip system is overhead sprinkling – that is, hanging the pipes above the level of the growing beds, in order to water the plants from above. This is something some plants prefer. The advantage of this strategy is that it leaves the surface of the soil free, and its disadvantage is that it increases the humidity levels more than the drip system. Like dripping pipes, sprinkling pipes can be hand-made by boring small holes along them, at regular intervals; however, in this case, the holes in the pipes should not be at their sides, but at their bottom, in order to avoid sprinkling the water away from the sides due to pressure.

4.8.3.4. *Mist systems*

Misting is the next level of sophistication with respect to sprinkling. It assures an even distribution of water in the air and on the plants and the growing soil, but it requires higher water pressures, which in turn may require additional or more powerful pumps, as well as sophisticated nozzles for misting.

4.8.3.5. *Flood floors (ebb-and-flow systems)*

Flooding, in a controlled fashion, a concrete floor onto which growth pots or containers have been placed is a relatively new technique of professional irrigation, become common in the 1990s. It has been observed by some researchers that the technique may entail the risk of spreading diseases and pests among plants, but the assessments performed seem to show that this risk is very low (Fisher 2003).

On references

The ebb-and-flood irrigation strategy is thoroughly presented in Fisher (2003).

4.9. Solutions for water catchment and storage suitable for self-building

The slope of a greenhouse roof can be devised to collect the rainwater and store it for watering the plants. The water can be stored with a seasonal cycle, in order to use in the dry season the water stored during the rainy season, or it can be stored for a shorter period of time, in order to use the water at a time not very far from the time of collection.

In the case of seasonal storage, the quantity of water to be stored is usually high, and the most affordable solution is that of storing it in a pool finished with a waterproof barrier on top – in cases of necessity, a folded thick polyethylene film – held in place by gravity. The extent of the waterproof films may determine the width of the pools (see Figure 4.5).

To avoid the use of pumps, the bottom level of the pools should be above the ground level in which the plants are embedded, in order to make it possible to distribute water by gravity. The pool may be built above the ground by constructing its edges with banks constituted by earth mounds. In this case, the flanks of the earth banks should be sloped, in order to resist the lateral thrust of the water due to gravity. The following section will specifically deal with this low-cost water storage solution.

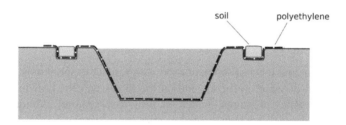

Figure 4.5. *Construction scheme for a low-cost pool for water collection. Steps: (1) a suitably shaped trench is dug in the ground. The slope of its walls should not be greater than that allowed by the shear resistance of the soil. (2) The surfaces of the pool should then be lined with a waterproofing layer. The lowest-cost solution for this layer is constituted by thick polyethylene (which has, however, short durability). (3) Long flaps should be kept at the edges of the waterproof lining, and should be blocked in position. A simple way of doing that is shown in the figure: it entails pressing the flaps down by placing each of them into a small trench in the ground, then filling it with soil*

4.9.1. *Creation of low-cost ponds*

In the ground below each earth bank, a small longitudinal trench should be dug, in order to avoid the bank sliding by one side under the thrust of the water. A small trench would also be needed at the top of the mound. This would allow the thick polyethylene film (which forms the waterproofing layer) to follow the contour of the mound thus obtained (see Figure 4.6). In addition, the trench space should be filled with earth, in order to keep the foil down.

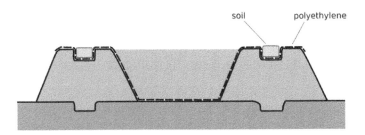

Figure 4.6. *A low-cost water pond can be constructed even without digging into the ground, by constructing the pool above the ground. This may be obtained by creating earth mounds of suitable height and length above the ground, adequate to constitute the banks of the pond. To avoid letting the mounds slide sideways on the ground under the thrust of the hydrostatic pressure of water, the mounds should be "locked" to the ground by means of "ground-locks", working by shape, as shown in the figure*

When the pools are not located above the ground level, but below it, the main water pool may be coupled with a short-term storage tank placed at a higher level, which should be filled at regular (commonly, daily) intervals by means of a pump, and from which the water for watering would be taken by gravity. This would spare the need to activate a pump each time some water is needed. In the described solution, the pump would only be used periodically, in order to fill the upper tank. This will make it possible to use a lower-cost pump, and reduce its maintenance requirements.

4.9.2. Rainwater collection

Rainwater collection is likely to be most useful in climates with alternated dry and humid seasons – for example, temperate climates, or subtropical climates. It is useful even in dry climates, but in those cases, the amount of water that can be collected from a greenhouse roof is likely to be largely insufficient – by itself – to the need of the crops in the greenhouse. Collecting water, instead, is likely not to be very useful in wet tropical climates, where rainwater is plentifully available.

The rainwater falling on the greenhouse can be collected at the ground level or at the level of the greenhouse gutters. The gutters may even be constituted by PVC films supported in such a way to be mildly sloped, and put in tension so as to disallow the creation of small local ponds that would prevent the water flow under the weight of the water itself.

When polyethylene films are used for building the water channel system (gutters and downpipes), as well as the flashings, care should be taken to substitute them periodically as an ordinary maintenance operation, because they would be subject to rapid decay due to solar radiation, as quickly as the polyethylene films enclosing the greenhouse.

Symmetric configurations of the water collection system constituted by a greenhouse with a symmetrically sloped roof – gable roof – and a pond (see Figure 4.7) have the drawback of making the water collection take place on both sides of the greenhouse unless the water is channeled from one side of the greenhouse to the other. To simplify the collection, asymmetrical roof configurations, like the one shown in Volume 3, Figures 3.30, 3.32, 3.33 and 3.34, may be useful, when their slope is sufficient to drive the water flow. On the contrary, in those asymmetric cases, the slope should be kept small enough to prevent the asymmetry from reducing the ventilation potential for some wind directions.

The water storage pool – and the possible additional polyethylene films aimed to channel the water – should be placed in such a way that the rainwater reaching the greenhouse roof flows to the pool by gravity (see Figure 4.7).

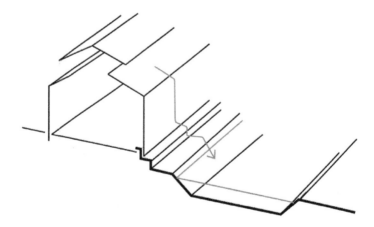

Figure 4.7. *By creating waterproofed continuity between the roof of a greenhouse and a pool, it is possible to collect the rainwater falling on the greenhouse roof and store it for the water requirements of the greenhouse itself. When the water pool is not dug into the ground, but raised above it, a pump may not be even needed to move the water into the greenhouse – gravity could be used. If the roof is gabled, like in this example, (a) one pond at each side of the greenhouse may be needed, or (b) the water at the side not served by the pond should be channeled at the ground level into the pond at the other side. The alternative, as seen, is constituted by an asymmetric, single-slope roof configuration*

The configurations characterized by continuity between roof and walls (like those in Volume 3, Figures 3.30, 3.31 and 3.33) have the advantage of making it possible for the water to flow from the roof to the wall without splashing. The configurations not featuring this continuity (e.g. those in Volume 3, Figures 3.32 and 3.34) have a higher probability of causing water spurts at the base of the greenhouse and make water collection more difficult. Even the configurations featuring clerestories (see Volume 3, Figure 3.29) can cause problems of water spurts – but smaller, because those would be localized on the roof.

Ideally, the water pools or tanks should be kept in shadow, so as to reduce the evaporation of water from them, which in hot climates can have a very high impact on the overall water balance of the system. An experimental solution that can be used in combination with the former for reducing water evaporation is placing a polyethylene film on the surface of water, and keeping it afloat by means of air-inflated tires.

On references

How-to references about water pond construction are Ludwig (2009) and Müller et al. (2015a).

Conclusion

The topics of structural frames and envelope systems (including frames) in greenhouse design are quite interrelated. Different transparent envelope systems can indeed be related to very different loads and require a very different amount of structural rigidity (and bracing criteria) from their supports – and therefore different structural schemes and rhythms. For example, although glass panels, polycarbonate panels and polyethylene films could in theory be installed on similar kinds of enclosures, the requirements they impose on the structures are usually so different that it is impossible, and even dangerously misleading, to design one without considering the other. And the opposite is also true: different types of greenhouse framing, characterized by different spans, stiffnesses and degrees of complexity, usually lead to – and are similar to – different kinds of enclosures.

What all these have in common is that the choices between different combinations of design options often depend on combinations of facts linked to the apparent randomness (and real complexity) of the world in which we live. In the specific case of wooden or timber frames, for example, a design solution may be largely dependent on which timber species and cuts are available at a given time.

This fact and myriad others remind us that more than one design option should always be made available, and that oversimplification – which also lurks behind the widely accepted distinction between the structure and the envelope – is always a danger.

For all of the aforementioned reasons, the contents of the second volume of this work (devoted to structural framing) and of the third one (devoted to enclosures) are meant to have a different relation between each other from those of the other two volumes (the first and the fourth). Indeed, the contents of these volumes are mutually intertwined up to the point of implying that they should be read and digested together. More than a closure, this conclusion is in fact a bridge towards the third volume. The two volumes together address the issue of greenhouse design for construction, in which the distinction between the structure and the envelope is mostly fictional, mostly descending from the objective of giving a shape to the construction "tale", rather than from the substance of reality.

References

AAVV (2002). *Wintergärten Und Gewächshäuser. Planung, Material, Werkzeug, Arbeitstechniken*. Moewig, Cologne.

Ahamed, M.S., Guo, H., Tanino, K. (2018). Development of a thermal model for simulation of supplemental heating requirements in Chinese-style solar greenhouses. *Computers and Electronics in Agriculture*, 150, 235–244. DOI:10.1016/j.compag.2018.04.025.

Albright, L.D., Aldrich, R.A., Both, A.-J., Graves, R.E., Howell Jr., J.C., Ross, D.S., Bartok Jr., J.W. (1978–2001). *Energy Conservation for Commercial Greenhouses*. Natural Resource, Agriculture, and Engineering Service (NRAES) Cooperative Extension, Ithaca, NY.

Aldrich, R.A. and Bartok, J.W. (1994). *Greenhouse Engineering*. Natural Resource, Agriculture, and Engineering Service (NRAES), Ithaca, NY.

Alinejad, T., Yaghoubi, M., Vadiee, A. (2020). Thermo-environomic assessment of an integrated greenhouse with an adjustable solar photovoltaic blind system. *Renewable Energy*, 156, 1–13. DOI:10.1016/j.renene.2020.04.070.

Aljubury, I.M.A. and Ridha, H.D. (2017). Enhancement of evaporative cooling system in a greenhouse using geothermal energy. *Renewable Energy*, 111, 321–331. DOI:10.1016/j.renene.2017.03.080.

Allen, E. (2014). *Fundamentals of Building Construction: Materials and Methods*, 6th edition. John Wiley & Sons, New York.

Allen, E. and Iano, J. (1992). *Architectural Detailing. Function, Constructibility, Aesthetics*, 2nd edition. John Wiley & Sons, New York.

Allen, E. and Iano, J. (2017). *The Architect's Studio Companion. Rules of Thumb for Preliminary Design*, 6th edition. Wiley, New York. Available at: http://archive.org/details/The_Architects_Studio_Companion_Rules_of_Thumb_for_Preliminary_Design_5th_Editio.

Allen, R.G., Pereira, L.S., Raes, D., Smith, M. (1998). *Crop Evapotranspiration – Guidelines for Computing Crop Water Requirements*. FAO irrigation and drainage paper 56, Rome.

Alward, R. and Shapiro, A. (1980). *Low-Cost Passive Solar Greenhouses. A Design and Construction Handbook*, 2nd edition. National Center for Appropriate Technology, Butte, MT.

Anderson, L.O. (1969). *Low-Cost Wood Homes for Rural America – Construction Manual*. Algrove Publishing, Almonte, Ontario [Online]. Available at: https://www.srs.fs.usda.gov/pubs/1130.

Angus-Butterworth, L.M. (1958). Glass. In *A History of Technology*, Volume 4, Singer, C., Holmyard, E.J., Hall, A.R., Williams, T.I. (eds). Oxford University Press, Oxford.

Asa'd, O., Ismet Ugursal, V., Ben-Abdallah, N. (2019). Investigation of the energetic performance of an attached solar greenhouse through monitoring and simulation. *Energy for Sustainable Development*, 53, 15–29. DOI:10.1016/j.esd.2019.09.001.

Aste, N. (2008). *Il fotovoltaico in architettura. L'integrazione dei sistemi per la generazione di elettricità solare*. Sistemi Editoriali, Naples.

Bambar, J. and Athienitis, A.K. (2019). Energy and economic analysis for the design of greenhouses with semi-transparent photovoltaic cladding. *Renewable Energy*, 131, 1274–1287. DOI:10.1016/j.renene.2018.08.020.

Bargach, M.N. and Boukallouch, D.M. (1999). A heating system using a flat plate collector to improve the inside greenhouse microclimate in Morocco. *Renewable Energy*, 18(3), 367–381.

Bastian, H.-W. (2000). *Wintergaerten planen un bauen*. Falken Verlag, Niederhausen.

Bazgaoua, A., Fatnassi, H., Bouharroud, R., Elame, F., Ezzaeri, K., Gourdo, L., Wifaya, A., Demrati, H., Tiskatine, R., Bekkaoui, A. et al. (2020). Performance assessment of combining rock-bed thermal energy storage and water-filled passive solar sleeves for heating Canarian greenhouse. *Solar Energy*, 198, 8–24. DOI:10.1016/j.solener.2020.01.041.

Ben Bonham, M. (2019). *Bioclimatic Double-Skin Facades*. Routledge, London.

Benoit, Y. (2014). *La maison à ossature bois par les schémas*. Eyrolles, Paris.

Benson, T. (1997). *The Timber-Frame Home. Design, Construction, Finishing*. Taunton Press, Newtown, CT.

Beytes, C. (ed.) (2003). *Ball Redbook. Vol. 1. Greenhouses and Equipment*, 17th edition. Ball Publishing, Batavia, IL.

Borer, P. and Harris, C. (1994–1997). *Out of the Woods. Ecological Design for Timber-Frame Housing*. Centre for Alternative Technology Publications, Machynlleth.

Bouadila, S., Lazaar, M., Skouri, S., Kooli, S., Farhat, A. (2014). Assessment of the greenhouse climate with a new packed-bed solar air heater at night, in Tunisia. *Renewable and Sustainable Energy Reviews*. 35, 31–41. DOI: 10.1016/j.rser.2014.03.051.

Bourgois, J. and Guyonnet, R. (1988). Characterization and analysis of torrefied wood. *Wood Science and Technology*, 22, 43–155.

Burrows, J. (1967–2013). *Canadian Wood-Frame House Construction*. Canada Mortgage and Housing Corporation, Ottawa.

Canadian Wood Council (1991). *Wood Reference Handbook: A Guide to the Architectural Use of Wood in Building Construction*. Canadian Wood Council, Ottawa.

Carmody, J., Selkowitz, S., Heschong, L. (1996). *Residential Windows. A Guide to New Technologies and Energy Performance*. W.W. Norton and Company, New York.

Carrol, J. (2005). *Working Alone: Tips & Techniques for Solo Building*. Taunton Press, Newtown, CT.

Carter, J. (ed.) (1981). *Solarizing Your Present Home: Practical Solar Heating Systems You Can Build*. Rodale Press, Emmaus, PA.

Chappel, S. (2011). *A Timber Framer's Workshop: Joinery & Design Essentials for Building Traditional Timber Frames*, 3rd edition. Fox Maple Press, Brownfield, ME.

Chen, C., Li, Y., Li, N., Wei, S., Yang, F., Ling, H., Yu, N., Han, F. (2018). A computational model to determine the optimal orientation for solar greenhouses located at different latitudes in China. *Solar Energy*, 165, 19–26. DOI:10.1016/j.solener.2018.02.022.

Clegg, P. and Watkins, D. (1978). *The Complete Greenhouse Book. Building and Using Greenhouses from Cold-Frames to Solar Structures*. Garden Way Publishing, Charlotte, VT.

Compagno, A. (2002). *Intelligent Glass Facades. Materials, Practice, Design*, 5th edition. Birkhauser, Basel.

Corkish, R. and Prasad, D. (2006). Integrated solar photovoltaics for buildings. *Journal of Green Building*, 1(2), 63–76. DOI:10.3992/jgb.1.2.63.

Cossu, M., Murgia, L., Ledda, L., Deligios, P.A., Sirigu, A., Chessa, F., Pazzona, A. (2014). Solar radiation distribution inside a greenhouse with south-oriented photovoltaic roofs and effects on crop productivity. *Applied Energy*, 133, 89–100. DOI:10.1016/j.apenergy.2014.07.070.

Cossu, M., Cossu, A., Deligios, P.A., Ledda, L., Li, Z., Fatnassi, H., Poncet, C., Yano, A. (2018). Assessment and comparison of the solar radiation distribution inside the main commercial photovoltaic greenhouse types in Europe. *Renewable and Sustainable Energy Reviews*, 94, 822–834. DOI:10.1016/j.rser.2018.06.001.

Cowan, H.J. (1971). *Architectural Structures: An Introduction to Structural Mechanics*. Pitman Publishing, London.

Cowan, H.J. (1980). *Solar Energy Applications in the Design of Buildings*. Applied Science Publisher, London.

Cowan, H.J. and Smith, P.R. (1983). *Environmental System*. Van Nostrand Reinhold, New York.

Cowan, H.J. and Wilson, F. (1981). *Structural Systems*. Van Nostrand Reinhold, New York.

Cuce, E., Harjunowibowo, D., Cuce, P.M. (2016). Renewable and sustainable energy-saving strategies for greenhouse systems: A comprehensive review. *Renewable and Sustainable Energy Reviews*, 64, 34–59. DOI:10.1016/j.rser.2016.05.077.

Dai, L., Li, S., Mu, L.D., Li, X., Shang, Y., Dong, M. (2015). Experimental performance analysis of a solar assisted ground source heat pump system under different heating operation modes. *Applied Thermal Engineering*, 75, 325–333. DOI:10.1016/j.applthermaleng.2014.09.061.

DeKay, A. and Brown, G.Z. (2014). *Sun, Wind and Light. Architectural Design Strategies*, 3rd edition. John Wiley & Sons, New York.

Drexel, T. (1999). *Wintergaerten bauen*. Callwey Verlag, Munich.

Eccli, E. (ed.) (1976). *Low-Cost, Energy Efficient Shelter for the Owner and Builder*. Rodale Press, Emmaus, PA.

von Elsner, B., Briassoulis, D., Waaijenberg, D., Mistriotis, A., von Zabeltitz, C., Gratraud, J., Russo, G., Suay-Cortes, R. (2000a). Review of structural and functional characteristics of greenhouses in European Union countries. Part I: Design requirements. *Journal of Agricultural Engineering Research*, 75, 1–16.

von Elsner, B., Briassoulis, D., Waaijenberg, D., Mistriotis, A., von Zabeltitz, C., Gratraud, J., Russo, G., Suay-Cortes, R. (2000b). Review of structural and functional characteristics of greenhouses in European Union countries. Part II: Typical designs. *Journal of Agricultural Engineering Research*, 75, 111–126. DOI: 10.1006/jaer.1999.0512.

Emmi, G., Zarrella, A., De Carli, M., Galgaro, A. (2015). An analysis of solar assisted ground source heat pumps in cold climates. *Energy Conversion and Management*, 106, 660–675. DOI: http://dx.doi.org/10.1016/j.enconman.2015.10.0160.

Engel, H. (1987). *Measure and Construction of the Japanese House*. Charles Tuttle, Rutland, VT.

Engel, A. (2011). *Carpentry Complete. Expert Advice from Start to Finish*. Taunton Press, Newtown, CT.

Esen, M. and Yuksel, T. (2013). Experimental evaluation of using various renewable energy sources for heating a greenhouse. *Energy and Buildings*, 65, 340–351. DOI:10.1016/j.enbuild.2013.06.018.

Feirer, J.L., Hutchings, G.R., Feirer, M.D. (1997). *Carpentry and Building Construction*, 5th edition. McGraw-Hill, New York.

Fisch, K. (2001). *Selbst Wintergärten und Glashäuser bauen*. Compact Verlag, Munich.

Fischer, R. and Yanda, B. (1977). *The Food and Heat Producing Solar Greenhouse. Design Construction Operation*. John Muir Publications, Costa Rica.

Fisher, R. (2003). Ebb-and-flood irrigation. In *Ball Redbook. Vol. 1. Greenhouses and Equipment*, 17th edition, Beytes, C. (ed.). Ball Publishing, Batavia, IL.

Ford, E.R. (2003a). *The Details of Modern Architecture. Vol. 1*. MIT Press, Cambridge, MA.

Ford, E.R. (2003b). *The Details of Modern Architecture. Vol. 2. 1928 to 1988*. MIT Press, Cambridge, MA.

Freeman, M. (1994). *How to Design and Build a House You Heat with the Sun*. Stackpole Books, Mechanicsburg, PA.

Freeman, M. (1997). *Build Your Own Greenhouse*. Stackpole Books, Mechanicsburg, PA.

Ganguly, A. and Ghosh, S. (2007). Modeling and analysis of a fan–pad ventilated floricultural greenhouse. *Energy and Buildings*, 39, 1092–1097. DOI:10.1016/j.enbuild.2006.12.003.

Gauld, B.J.B. (1984). *Structures for Architects*, 3rd edition. Routledge, London.

Ghosal, M.K. and Tiwari, G.N. (2006). Modeling and parametric studies for thermal performance of an earth to air heat exchanger integrated with a greenhouse. *Energy Conversion and Management*, 47, 1779–1798. DOI:10.1016/j.enconman.2005.10.001.

Ghosal, M.K., Tiwari, G.N., Srivastava, N.S.L. (2004). Thermal modeling of a greenhouse with an integrated earth to air heat exchanger: An experimental validation. *Energy and Buildings*, 36(3), 219–227. DOI:10.1016/j.enbuild.2003.10.006.

Ghosal, M.K., Tiwari, G.N., Das, D.K., Pandey, K.P. (2005). Modeling and comparative thermal performance of ground air collector and earth air heat exchanger for heating of greenhouse. *Energy and Buildings*, 37(6), 613–621. DOI:10.1016/j.enbuild.2004.09.004.

Ghosh, A. (2020). *Greenhouse Technology. Principles and Practice*. CRC Press, Boca Raton, FL.

Gordon, J.E. (1978). *Structures: Or Why Things Don't Fall Down*. Pelican Books, London.

Gourdo, L., Fatnassi, H., Tiskatine, R., Wifaya, A., Demrati, H., Aharoune, A., Bouirden, L. (2019). Solar energy storing rock-bed to heat an agricultural greenhouse. *Energy*, 169, 206–212. DOI:10.1016/j.energy.2018.12.036.

Grondzik, W.T., Kwok, A.G., Stein, B., Reynolds, J.S. (2019). *Mechanical and Electrical Equipment for Buildings*, 13th edition. John Wiley & Sons, New York.

Gulrez, S.K.H., Abdel-Ghany, A.M., Al-Helal, I.M., Al-Zaharani, S.M., Alsadon, A.A. (2013). Evaluation of PE films having NIR-reflective additives for greenhouse applications in arid regions. *Advances in Materials Science and Engineering*, DOI:10.1155/2013/575081.

Hait, J. (1985). Passive annual heat storage: Improving the design of earth sheltered homes. *Mother Earth News*, January/February.

Hait, J. (2013). *Passive Annual Heat Storage. Improving the Design of Earth Shelters*, 3rd edition. Rocky Mountain Research Center, Fort Collins, CO.

Hanan, J.J. (1997). *Advanced Technology for Protected Environment*. CRC Press, Boca Raton, FL.

Hanan, J.J. (1998). *Advanced Technology for Protected Horticulture*. CRC Press, Boca Racon, FL.

Haun, L. (1998). *The Very Efficient Carpenter*. Taunton Press, Newtown, CT.

Haupt, E. (2001). *Wintergaerten und Glasanbauten im Detail*. WEKA Baufachverlag, Augsburg.

Haupt, E. and Wiktorin, A. (1996). *Wintergaerten. Ein Praxis-Handbuch*. Oekobuch Verlag, Freiburg.

Hayes, J. and Gillett, D. (1977). *Proceedings of the Conference of Energy-Conserving, Solar-Heated Greenhouses*. Marlboro College, Marlboro, VT.

Hegger, M., Fuchs, M., Stark, T., Zeurner, M. (2008). *Energy Manual. Sustainable Architecture*. Birkhaeuser, Basel.

Hematian, A., Ajabshirchi, Y., Ranjbar, S.F., Taki, M. (2019). An experimental analysis of a solar-assisted heat pump (SAHP) system for heating a semisolar greenhouse. *Energy Sources, Part A: Recovery, Utilization, and Environmental Effects*. DOI:10.1080/15567036.2019.1663308.

Hens, H. (2012a). *Performance Based Building Design 1. From Below Grade Construction to Cavity Walls*. Ernst and Sohn, Berlin.

Hens, H. (2012b). *Building Physics. Heat, Air and Moisture. Fundamental and Engineering Methods with Examples and Exercises*. Ernst and Sohn, Berlin.

Hens, H. (2012c). *Performance-based Building Design 2. From Timber Frame Construction to Partition Walls*. Ernst und Sons, Berlin.

Herzog, T. and Natterer, J. (1984). *Habiller de verre et de bois. Agrandir des maisons familiales sans augmenter la consommation d'energie*. Presses Polytechniques Romandes, Lausanne.

Herzog, T., Natterer, J., Schweitzer, R., Volz, M., Winter, W. (2004). *Timber Construction Manual*. Birkhäuser, Basel [original edition: Holzbau Atlas, 1991].

Hyde, R. (2000). *Climate Responsive Design. A Study of Buildings in Moderate and Hot Humid Climates*. E and FN Spon, London.

Ibbotson, H. (1964). *Build Your Own Greenhouse*. Foyle, London.

IEA (1999). *Solar Air Systems. Built Examples*. James and James, London.

IEA (2000). *Solar Air Systems. A Design Handbook*. James and James, London.

Izard, J.L. (1993). *Architectures d'été. Construire pour le comfort d'été*. Édisud, Aix-en-Provence.

Jackson, A. and Day, D. (1986). *Complete Do It Yourself Manual*. HarperCollins, New York.

Jain, D. and Tiwari, G.N. (2002). Modeling and optimal design of evaporative cooling system in controlled environment greenhouse. *Energy Conversion and Management*, 43(16), 2235–2250. DOI: 10.1016/S0196-8904(01)00151-0.

Jain, D. and Tiwari, G.N. (2003). Modeling and optimal design of ground air collector for heating in controlled environment greenhouse. *Energy Conversion and Management*, 44(8), 1357–1372. DOI:10.1016/S0196-8904(02)00118-8.

Jeni, K. (2005). *Wintergärten und Glasanbauten: Planen, Bauen, Wohnen: Ideen und Beispiele mit Glas*. Blottner Fachverlag, Taunussstein.

Jones, T.H. (1978). *How to Build Greenhouses, Garden Shelters and Sheds*. Harper and Row, New York.

Jones, R.W. and McFarland, R.D. (1984). *The Sunspace Primer. A Guide to Passive Solar Heating*. Van Nostrand Reinhold, New York.

Kachadorian, J. (1997). *The Passive Solar House. Using Solar Design to Heat and Cool Your Home*. Chelsea Green Publishing, White River Junction, VT.

Kern, K. (1975). *The Owner-Built Home*. Scribner, New York.

Kern, B. and Kern, K. (1974). *The Owner-Built Homestead*. Charles Scribner's Sons, New York.

Kern, B. and Kern, K. (1981). *The Owner-Built Pole Frame House*. Charles Scribner's Sons, New York.

Kinkade-Levario, H. (2007) *Design for Water*. New Society Publishers, Gabriola Island.

Kolb, B. (1990). *Sonnenklar solar! Die neue Generation von Sonnenhaeusern*. Blok Verlag, Munich.

Kolb, J. (2008). *Systems in Timber Engineering: Loadbearing Structures and Component Layers*. Birkhauser, Basel.

Kornher, S. and Zaugg, A. (1984–2006). *The Complete Handbook of Solar Air Heating Systems. How to Design and Build Efficient, Economical Systems for Heating Your Home*. Rodale Press, Emmaus, PA.

Kürklü, A., Bilgin, S., Özkan, B. (2003). A study on the solar energy storing rock-bed to heat a polyethylene tunnel type greenhouse. *Renewable Energy*, 28, 683–697.

Kurth, H. and Kurth, G. (1982). *Greenhouses for Longer Summers*. Batsford, London.

Kwok, A.G. and Grondzik, W.T. (2007). *The Green Studio Handbook*. Elsevier, Amsterdam.

Lagier, H. and Dastot, F. (2008). *Les vérandas: conception, construction, entretien maintenance*. CSTB, Marne-la-Vallée.

Lazaar, M., Bouadila, S., Kooli, S., Farhat, A. (2015). Comparative study of conventional and solar heating systems under tunnel Tunisian greenhouses: Thermal performance and economic analysis. *Solar Energy*, 120, 620–635. DOI:10.1016/j.solener.2015.08.014.

Lechner, N. (2008). *Heating, Cooling. Lighting. Sustainable Design Methods for Architects*, 3rd edition. John Wiley & Sons, New York.

Lees, A. and Heyn, E.V. (1991). *Decks and Sunspaces*. Sterling Publishing Company Inc, New York.

Li, B., Shi, B., Yao, Z., Shukla, M.K., Du, T., (2020). Energy partitioning and microclimate of solar greenhouse under drip and furrow irrigation systems. *Agricultural Water Management*, 234. DOI: 10.1016/j.agwat.2020.106096.

Lippsmeier, G. (1980). *Tropenbau / Building in the Tropics*. Callwey, Munich.

Liu, Z., Wu, D., He, B.-J., Liu, Y., Zhang, X., Yu, H., Jin, G. (2018). Using solar house to alleviate energy poverty of rural Qinghai-Tibet region, China: A case study of a novel hybrid heating system. *Energy and Buildings*, 178, 294–303. DOI:10.1016/j.enbuild.2018.08.042.

Lu, W., Zhang, Y., Fang, H., Ke, X., Yang, Q. (2017). Modelling and experimental verification of the thermal performance of an active solar heat storage-release system in a Chinese solar greenhouse. *Biosystems Engineering*, 160, 12–24. DOI:0.1016/j.biosystemseng.2017.05.006.

Ludwig, A. (2009). *Water Storage. Tanks, Cisterns, Acquifers, and Ponds*, 2nd edition. Oasis Design, Santa Barbara, CA.

Marshall, R. (2006). *How to Build Your Own Greenhouse*. Storey Publishing, North Adams, MA.

Matana, M. (1999). *Vérandas*. CME, Ashford, Kent.

Matuska, T. and Sourek, B. (2017). Performance analysis of photovoltaic water heating system. *International Journal of Photoenergy*. DOI:10.1155/2017/7540250.

Mauldin, J.H. (1987). *Sunspaces. Home Additions for Year-Round Natural Living*. Tab Books, Blue Ridge Summit, PA.

Mazria, E. (1979). *The Passive Solar Energy Book*. Rodale Press, Emmaus, PA.

McCartney, L., Orsat, V., Lefsrud, M.G. (2018). An experimental study of the cooling performanceand airflow patterns in a model Natural Ventilation Augmented Cooling (NVAC) greenhouse. *Biosystem Engineering*, 74, 173–189. DOI:10.1016/j.biosystemseng.2018.07.005.

McCullagh, J.C. (1978). *The Solar Greenhouse Book*. Rodale Press, Emmaus, PA.

Miller, C. (2004). *Building Tips and Techniques*. Taunton Press, Newtown, CT.

Mincke, G. and Mahlke, F. (2005). *Building with Straw*. Birkhäuser, Basel.

Mollison, B.C. and Slay, R.M. (1991). *Introduction to Permaculture*. Tagari Publications, Tasmania.

Moro, J.L., Rottner, M., Alihodzic, B., Weissback, M. (2009). *Baukonstruktion vom Prinzip zu Detail – Band 3 – Umsetzung*. Springer, Berlin.

Müller, A., Kariuki, I., Jokisch, A. (2015a). *CuveWaters Construction Manual # 3 – Pond*. Technische Universität Darmstadt, Darmstadt [Online]. Available at: http://www.cuvewaters.net/Publications.

Müller, A., Kariuki, I., Jokisch, A., Witkowski, R. (2015b). *CuveWaters Construction Manual # 5 – Drip Irrigation*. Technische Universität Darmstadt, Darmstadt [Online]. Available at: http://www.cuvewaters.net/Publications.

National Research Council of Canada (1981). *The Solarium Workbook*. National Research Council of Canada, Ottawa.

Neal, C.D. (1975). *Build Your Own Greenhouse: How to Construct, Equip, and Maintain It*. Chilton Book Co, Radnor, PA.

Nengelken, P.H. (1996). *Wintergarten un Ueberdachungen. Plannen, Bauen, Bepflanzen*. BLV, Zurich.

New York State Energy Research and Development Authority (1982). *Making Your Own Solar Wall Panel. Complete Instructions and Plans for Constructing Three Different Solar Wall Panel Systems*. New York.

Oesterle, E., Lieb, R.-D., Lutz, M., Heusler, W. (2001). *Double-Skin Facades. Integrated Planning*. Prestel, Munich.

Ouazzani Chahidi, L., Fossa, M., Priarone, A., Mechaqrane, A. (2021). Energy saving strategies in sustainable greenhouse cultivation in the Mediterranean climate – A case study. *Applied Energy*, 282. DOI:10.1016/j.apenergy.2020.116156.

Ozgener, L. (2011). A review on the experimental and analytical analysis of earth to air heat exchanger (EAHE) systems in Turkey. *Renewable and Sustainable Energy Reviews*, 15, 4483–4490. DOI:10.1016/j.rser.2011.07.103.

Ozgener, L. and Ozgener, O. (2010a). An experimental study of the exergetic performance of an underground air tunnel system for greenhouse cooling. *Renewable Energy*, 35(2010), 2804–2811. DOI:10.1016/j.renene.2010.04.038.

Ozgener, L. and Ozgener, O. (2010b). Energetic performance test of an underground air tunnel system for greenhouse heating. *Energy*, 35, 4079–4085. DOI:10.1016/j.energy.2010.06.020.

Ozgenera, O. and Hepbasli, A. (2005). Performance analysis of a solar-assisted ground-source heat pump system for greenhouse heating: An experimental study. *Building and Environment*, 40(8), 1040–1050. DOI:10.1016/j.buildenv.2004.08.030.

Peng, Z., Deng, W., Tenorio, R. (2020). An integrated low-energy ventilation system to improve indoor environment performance of school buildings in the cold climate zone of China. *Building and Environment*, 182. DOI:10.1016/j.buildenv.2020.107153.

Pierce, J. (1981–1992). *Home Solar Gardening*. Key Porters Book Limited, Toronto.

Pracht, K. (1984). *Holzbau-Systeme*. Rudolf Müller, Cologne.

Priavolou, C. and Niaros, V. (2019). Assessing the openness and conviviality of OpenSource technology: The case of the WikiHouse. *Sustainability*, 11(17), 4746. DOI: 10.3390/su11174746.

Price, P. and Greer, N.R. (2009). *Greenhouses and Garden Sheds*. Creative Publishing International, Minneapolis, MN.

Ren, J., Zhao, Z., Zhang, J., Wang, J., Guo, S., Sun, J. (2019). Study on the hygrothermal properties of a Chinese solar greenhouse with a straw block north wall. *Energy and Buildings*, 193, 127–138. DOI:10.1016/j.enbuild.2019.03.040.

Roy, R. (2004). *Timber Framing for the Rest of Us. A Guide to Contemporary Post and Beam Construction*. New Society Publishers, Gabriola Island.

Salvadori, M. and Heller, R. (1963). *Structure in Architecture*. Prentice-Hall, Englewood Cliffs, NJ.

Schiller, L. and Plincke, M. (2016). *The Year-Round Solar Greenhouse. How to Design and Build a Net-Zero Energy Greenhouse*. New Society Publishers, Gabriola Island.

Schmidt, P. (2011). *The Complete Guide to Greenhouses and Garden Projects. Greenhouses, Cold Frames, Compost Bins, Trellises, Planting Beds, Potting Benches, and More*. Creative Publishing International, Minneapolis, MN.

Schodek, D. and Bechthold, M. (2013). *Structures*, 7th edition. Pearson, London.

Schwolsky, R. and Williams, J.I. (1982). *The Builder's Guide to Solar Construction*. McGraw-Hill, New York.

Scudo, G. (1993). *Tecnologie termodilizie. Principi e tecniche innovative per la climatizzazione dell'edilizia*. CittàStudi, Milan.

Sethi, V.P., Sumathy, K., Lee, C., Pal, D.S. (2013). Thermal modeling aspects of solar greenhouse microclimate control: A review on heating technologies. *Solar Energy*, 96, 56–82. DOI:10.1016/j.solener.2013.06.034.

Shan, M., Yu, T., Yang, X. (2016). Assessment of an integrated active solar and air-source heat pump water heating system operated within a passive house in a cold climate zone. *Renewable Energy*, 87, 1059–1066. DOI:10.1016/j.renene.2015.09.024.

Shapiro, A.M. (1984). *The Homeowner's Complete Handbook for Add-On Solar Greenhouses and Sunspaces: Planning, Design, Construction*. Rodale Press, Emmaus, PA.

Shukla, A., Tiwari, G.N., Sodha, M.S. (1987). Analysis of thermal performance of a greenhouse as a solar collector. *Energy in Agriculture*, 6(1), 17–26. DOI:10.1016/0167-5826(87)90018-0.

Shurcliff, W.A. (1979). *New Inventions in Low-Cost Solar Heating*. Brick House Publishing, Andover, MA.

Smith, S. (1992). *Greenhouse Gardener Companion*. Fulcrum Publishing, Golden, CO.

Smith, E. (2011). *DIY Solar Projects*. Creative Publishing International, Minneapolis, MN.

Sobon, J.A. and Schroeder, R. (1984). *Timber Frame Construction. All About Post-and-Beam Building*. Storey Publishing, North Adams, MA.

Spence, W.P. (1999). *Carpentry & Building Construction: A Do-It-Yourself Guide*. Sterling Publishing Company, New York.

Stempel, U.E. (2008). *Wintergarten selbst planen und bauen*. Franzis Verlag, Munich.

Steven Winter Associates (1983). *The Passive Solar Design and Construction Handbook*. John Wiley & Sons, New York.

Szokolay, S.V. (1975). *Solar Energy and Buildings*. The Architectural Press, London.

Taylor, T.M. (1998). *Secrets to a Successful Greenhouse and Business*. GreenEarth Publishing Company, Melbourne, FL.

Temple, P.L. and Adams, J.A. (1981). *Solar Heating: A Construction Manual*. Total Environmental Action, Harrisville, NH.

Thallon, R. (2016). *Graphic Guide to Frame Construction*, 4th edition. Taunton Press, Newtown, CT.

Timber Engineering Company (1956). *Timber Design and Construction Handbook*. McGraw-Hill, New York.

Timberlake, J. and Kieran, S. (2008). *Loblolly House: Elements of a New Architecture*. Princeton Architectural Press, Princeton, NJ.

Timm, U. (2000). *Der Wintergaerten Wohnraume unter Glas*. Callwey, Munich.

Tiwari, G.N., Dubey, A.K., Goyal, R.K. (1997). Analytical study of an active winter greenhouse. *Energy*, 22(4), 389–392.

Tiwari, G.N., Akhtar, M.A., Shukla, A., Emran Khan, M. (2006). Annual thermal performance of greenhouse with an earth–air heat exchanger: An experimental validation. *Renewable Energy*, 31, 2432–2446. DOI:10.1016/j.renene.2005.11.006.

Toht, D. (2013). *40 Project for Building Your Backyard Homestead. A Hand-On, Step-by-Step Sustainable Living Guide*. Creative Homeowner, Upper Saddle River, NJ.

Torroja, E. (2011). *The Structures of Eduardo Torroja: An Autobiography of an Engineering Accomplishment*. Literary Licensing, Whitefish, MT.

Total Environmental Action and Los Alamos Narural Laboratory (1984). *Passive Solar Design Handbook*. Van Nostrand Rheinhold, New York.

Twitchell, M. (1985). *Solar Projects for Under $500*. Garden Way Publishing, Charlotte, VT.

UNHCR (2006). Large ferro-cement water tank design parameters and construction details [Online]. Available at: http://www.unhcr.org/publications/operations/49d089a62/large-ferro-cement-water-tank-design-parameters-construction-details.html.

Vitruvius, Marcus Pollio (1998). *Vitruvius: The Ten Books on Architecture*. Dover Publications, New York.

Wade, A. (1980). *A Design and Construction Handbook for Energy-Saving Houses*. Rodale Press, Emmaus, PA.

Wade, A. and Ewenstein, N. (1977). *30 Energy Efficient Houses... You Can Build*. Rodale Press, Emmaus, PA.

Wagner, J.D. (2005). *House Framing*. Creative Homeowner, Upper Saddle River, NJ.

Wagner, J.D. and DeKorne, C. (2002). *Barns, Sheds, and Outbuildings*. Creative Homeowner, Upper Saddle River, NJ.

Watson, D. (1977). *Designing and Building a Solar House*. Garden Way Publishing, Charlotte, VT.

Watson, D. and Labs, K. (1983). *Climatic Building Design. Energy-Efficient Building Principles and Practice*. McGraw-Hill, New York.

Weichun, T., Minjun, S., Xuetao, Z. (2008). Water-saving technologies, solar green house and ecological rehabilitation in Minqin Oasis of Gansu Province, China. *Chinese Journal of Population Resources and Environment*, 6(3), 73–82. DOI:10.1080/10042857.2008.10684885.

Williams, J. (1980). *How to Build and Use Greenhouses*. Ortho Books, San Francisco, CA.

Wing, C. (1990). *Visual Handbook of Building and Remodeling*. Rodale Press, Emmaus, PA.

Winter Associates (1983). *Passive Solar Construction Handbook*. Rodale Press, Emmaus, PA.

Wolfe, R., Merrilees, D., Loveday, E. (1980). *Low-Cost Pole Building Construction*. Storey Books, North Adams, MA.

Wolpert, C.R. (1989). *Sun Rooms*. Price Stern Sloan, Los Angeles, CA.

Wright, F.L. (1954). *The Natural House*. Horizon Press, New York.

von Zabelitz, C. and Boudoin, W.O. (2005). *Greenhouses and Shelter Structures for Tropical Climates*. Daya Publishing House, New Delhi and FAO, Rome.

Zappone, C. (2009). *La serra solare*, 2nd edition. Sistemi Editoriali, Napoli.

Zaragoza, G., Buchholz, M., Jochum, P., Pérez-Parra, J. (2007). Watergy project: Towards a rational use of water in greenhouse agriculture and sustainable architecture. *Desalination*, 211, 296–303. DOI:10.1016/j.desal.2006.03.599.

Zhang, X., Lv, J., Dawuda, M.M., Xie, J., Yu, J., Gan, Y., Zhang, J., Tang, Z., Li, J. (2019). Innovative passive heat-storage walls improve thermal performance and energy efficiency in Chinese solar greenhouses for non-arable lands. *Solar Energy*, 190, 561–575. DOI:10.1016/j.solener.2019.08.056.

Zhang, K., Yu, J., Ren, Y. (2022). Research on the size optimization of photovoltaic panels and integrated application with Chinese solar greenhouses. *Renewable Energy*, 182. DOI:10.1016/j.renene.2021.10.031.

Zhou, N., Yu, Y., Yi, J., Liu, R. (2017). A study on thermal calculation method for a plastic greenhouse with solar energy storage and heating. *Solar Energy*, 142, 39–48. DOI:10.1016/j.solener.2016.12.016.

Index

A, B

air
 channels, 192–194
 pipes, 192, 195, 200, 211
bacteria, 132, 133, 200
beams, 2–5, 12, 14–17, 20, 24–27, 37, 39,
 42–46, 51, 61, 63, 64, 67–72, 76, 79,
 80, 82, 84–86, 88–99, 105, 110–130,
 132, 136, 137, 139, 145, 148, 149, 152,
 157, 158, 160, 172, 184, 188, 189
 composite, 96
 continuous, 84, 90–94, 137
 double, 71, 90, 93–95, 97, 111, 117,
  125, 145
 fascia (*see also* header joists), 69, 70
 post-and-beam, 46, 111, 113, 118
 primary, 97, 110, 111, 118, 129,
  130, 188
 sloped, 63, 112–117
birdsmouth cuts, 19
boilers, 192, 193, 205, 207
bolts, 2, 6, 9, 11, 12, 14, 20, 48, 50, 77,
 79, 82–94, 96, 100–109, 115, 119, 120,
 126, 128, 137, 139, 144, 152
bracing, 2, 6, 39, 46–51, 56, 61, 63,
 82, 83, 118, 120, 127, 149–157, 159,
 160, 163, 225

C, D

cables, 1, 47, 49, 50, 52, 55, 56
Canadian wells, 194, 197, 198
cantilevers, 25–27, 33, 39, 54
charring, 133, 175
chillers, 212
chords, 44–46
columns, 1, 45, 46, 51, 61, 63, 64, 68, 76,
 79, 82, 84–86, 88–97, 99, 101, 103,
 105, 111, 116, 119, 121, 123, 126, 127,
 132, 152
 continuous, 84, 90–93
 double, 90, 93–95, 97
connections
 bolted, 14
 glued, 15
 half-wood, 49, 98, 153–155, 157,
  158
 nailed, 12, 13, 85, 86, 91, 106
 screwed, 6, 13, 14, 106, 107
connectors
 ring, 48, 87, 91, 92, 106, 111,
  128
 shear, 91, 93, 103, 104, 106, 119, 137,
  139
corbels, 77, 79, 84, 93, 94, 115, 117, 137,
 139, 144

diagonals, 82, 83, 119, 120, 124, 126, 127, 149, 150, 152–158, 160
domes, 54, 57–59
   geodesic, 57–59

E, F

earth mounds, 220, 221
evaporative cooling pads, 212, 213
façades
   front, 69, 71, 110, 111, 117, 118, 120, 130–132
   gable, 131
fans, 192–195, 198, 206–209, 212–214
   coils, 192, 209
floors, 168, 184, 187–189, 192, 200, 202, 206, 210, 219
   platform, 141
   platform-frame, 3, 5, 6, 15, 51, 148, 188
foundations, 2–4, 10, 21, 32, 33, 39, 41, 51, 57, 98, 100, 101, 103, 112, 118, 132, 133, 138–142, 144, 147–149, 161
   flashing, 165, 167, 186
   insulation, 165–168, 184–186
   pads, 118
   pier, 184
   platform-on-poles, 148
   pole, 132–142, 146–149
      -and-trench, 148, 149
   sills, 2, 98, 161, 186
   stake-based, 172–174
   timber, 173, 175–178, 184, 188
      log and gravel trench, 175
   trench, 165, 166, 169, 170, 175–179, 181, 185, 186
   walls, 161–172, 176, 178, 184–187
      boulders-and-mortar, 169
      brick masonry, 170
      concrete, 172, 175, 176, 184, 187, 188
      hollow concrete block, 170, 172, 200
         parged, 172

   water drainage, 136, 170, 187
   wooden-frame, 172
   waterproofing, 164, 165, 168, 176, 178, 186
frames
   balloon, 2, 6
   light, 1–4, 6, 7, 9–12, 14, 15, 19–22, 28, 29, 32, 33, 42–53
   platform, 63, 111, 148, 188
   portal, 42–46, 48
   timber, 1, 3, 6, 7, 9, 12, 16, 32, 33, 37, 43, 44, 47, 51, 52
   trussed, 43

G, H

gables, 3, 20–28, 31, 34, 36
gaskets, 6, 25, 28
girders, 95, 136, 137, 139, 141, 143–146, 148, 149
gravel, 165, 166, 169–172, 175–179, 181, 185–188, 199, 211
   beds, 175, 187, 211
ground
   air heat transfer (GAHT) systems, 194–196
   pipes, 196, 197, 211
   thermal exchange, 194
gutters, 70, 112
header joists (*see also* fascia beams), 2, 5, 15–17, 25
heat exchangers
   air-to-air, 193, 194, 198, 212
   air-to-ground, 192, 194, 195, 202
   convective, 171, 200, 211, 212
   ground-to-air, 203, 205
   pumps, 192, 193, 197, 205, 206, 209–211, 214
   radiant, 212
   recovery, 193, 194
heating and cooling
   active
      cooling, 211, 212
      heating, 194, 196

air
  cooling, 207
  heating, 206
auxiliary
  cooling, 211
  heating, 204, 205
electric
  heating, 206, 207, 210
gas
  heating, 210
plants
  cooling, 191
  heating, 205, 210
radiant
  cooling, 212
  heating, 193, 209, 210, 212
radiators
  heating, 193, 212
systems
  air, 191–193, 209, 214
  water, 193, 209, 210
hollow-block channels, 201, 202

I, J, L

infiltration, 16, 20, 30
inserts, 83, 84, 105
joints
  butt, 7–11, 17, 43–46, 49, 56–58, 87, 94, 99, 105, 136, 152
  lap, 7, 11, 12, 15, 41, 43, 44, 49, 89–91, 93–95, 98, 106, 115, 119, 120, 123, 152, 155
  miter, 77
  scarf, 98–100
  tenon-and-mortice, 75–84, 120
    dovetail, 77–79
    open, 77–81
  vertical, 20, 30
ledgers, 3, 4, 16, 19, 35
lumber
  glued-laminated, 99, 110, 111, 129
  heat-treated, 134
  pressure-treated, 100, 133, 141, 162–164, 167, 173
  roasted, 134

M, N, O

mullions (*see also* studs), 1, 2, 5, 6, 16, 20–25, 39, 42, 48, 51, 61, 63, 64, 68, 70–72, 117, 121, 122, 137, 147
nails, 77, 79, 84, 87, 89–91, 94, 101, 105–109, 115, 119, 120, 128, 152
opaque enclosures, 29, 30, 47
operable openings, 30, 39

P, R

paints, 103, 133
panels (*see also* passive solar water and air panels, and photovoltaic (PV) panels), 1, 2, 6, 16, 17, 19, 21, 22, 25, 27–29, 32, 33, 42, 44, 45, 47, 48, 50, 51, 56, 57, 59
passive solar water and air panels, 166, 185, 191, 192, 194–196, 201, 205, 206, 210, 212, 214, 215
pavements, 187–189, 192, 193, 210, 213, 216
pegs, 79–83, 152
photovoltaic (PV) panels, 193, 205, 206, 210, 211, 213, 214
plates, 2–10, 13–18, 20–25, 44, 46, 48, 50, 57, 58, 70, 76, 77, 83–89, 92–98, 100, 101, 103, 104, 109, 112, 113, 119, 124, 133, 152
  cover, 84–87, 89, 94–96, 152
  tooth, 13, 14, 44, 46, 87, 91, 92, 119, 131, 152
poles, 118, 132–142, 146–149
  construction, 132, 133, 136–138, 146, 147, 149
posts, 61, 63, 64, 69, 71, 74, 79, 80, 82, 84, 100–105, 111–120, 128, 130, 132, 136, 138, 141–145
pots, 188, 215, 219
preservatives, 133
purlins, 63, 64, 71–73, 121, 122, 124, 125, 129, 137, 146, 154, 155, 157, 158

rafters, 1–5, 12, 16, 17–19, 21, 24–27, 37, 42, 43, 45, 46, 48, 51, 52
   edge, 17–19, 21, 24–27
rainscreens, 27, 30
renewable energy sources, 210
rock-beds, 200–202
rods, 13, 20, 47, 49, 50, 56
roofs, 2, 4, 15–17, 19–27, 31, 35, 36, 39, 42, 46, 47, 51, 53, 54, 57, 63, 67, 68, 70–73, 87, 111, 112, 116, 117, 121, 125, 130–132, 149, 151

S, T

screws, 77, 79, 84, 87, 89–91, 94, 101, 105–109, 115, 119, 120, 126, 128, 152
seals, 16, 20, 22, 27, 56, 57
shear, 61, 77, 79, 81, 84, 91–93, 103, 104, 106, 110, 111, 119, 120, 128, 137, 139, 152
span, 61, 63, 64, 67, 68, 72, 76, 90, 93, 94, 96, 97, 110, 111, 115, 116, 129, 130, 132, 133, 145, 149, 153
stack effect, 198, 199, 202
stakes, 135
steel
   angles, 84, 85
   footings, 100, 101, 104
   rods, 170–172, 175, 177
stoves, 209
stress, 61, 76, 84, 86, 87, 90, 96, 98, 101, 103, 106–110, 115, 119, 124, 151, 152, 168
structures
   reciprocal, 54
   trussed light-frame, 43, 123, 153

studs (*see also* mullions), 1, 2, 4–6, 10, 12, 14, 16, 17, 19–25, 28–30, 39, 42, 43, 48, 49, 51, 52, 61, 63, 64, 71, 72, 100, 117, 121, 137, 147
transoms, 3, 16, 17, 19, 21, 25, 31, 42, 63, 64, 68, 71, 73, 121, 122, 137
trusses, 14, 41, 44, 45, 48

U, V, W

uprights, 111, 119–121, 123–129, 151
vaults, 53–57
   lamella, 53–57
Venlo (greenhouse type), 131
walls
   back, 4, 19, 25, 30–34, 36–39, 41, 74, 75
   foundation, 98, 100, 101, 112, 133, 148
   gable, 3, 20–28, 34, 36
   shared, 19, 39, 41, 111
water
   catchment, 219
   containers, 216–218
   distribution, 216–218
      drip system, 218, 219
      ebb-and-flow systems, 219
      flood floors, 219
      mist system, 219
      sprinkling system, 219
   ponds, 220–223
   rain, 169, 186, 216, 217, 219, 221, 222
   storage, 200, 201, 217, 219–222
   tap, 216
watering systems, 191, 215
wood
   grain, 76, 78, 86, 90, 106, 107, 109, 110
   moisture-resistant, 132

Summary of Volume 1

Foreword

Remo DORIGATI

Introduction

Chapter 1. Basic Concepts

    1.1. What a greenhouse usually is – and what it could be
    1.2. The historical trajectory of greenhouses
    1.3. The main design factors: shape, orientation and envelope characteristics, in the context of local microclimates
        1.3.1. Climate analysis
        1.3.2. Site analysis
    1.4. Solar gains and air retention as conditions for the greenhouse effect
    1.5. Solar gains and thermal losses
        1.5.1. Facts common to all kinds of solar gains
        1.5.2. Factors influencing the solar gains on surfaces
    1.6. Thermal storage
        1.6.1. Charging the thermal masses by direct radiation
        1.6.2. Loading the thermal masses by reflected radiation
        1.6.3. Loading the thermal masses by convection
        1.6.4. Phase-change materials
        1.6.5. Natural convection – thermosyphoning
        1.6.6. Further information on the solar utilization of directly radiated thermal masses
    1.7. Passive ventilative cooling
        1.7.1. Indoor air movements
        1.7.2. Thermal buoyancy ventilation

1.7.3. Sound absorption for sound insulation when combined with ventilation strategies
1.7.4. Quantity of air changes
1.8. Dissipation of heat towards the sky
1.9. Dependence of solar control on the radiation type
1.9.1. Techniques and indicators for checking solar access
1.9.2. General shading strategies
1.9.3. Horizontal shading devices
1.9.4. Vertical shading devices
1.9.5. Horizontal shading devices for east and west exposures
1.9.6. Horizontal shading devices for south-east and south-west exposures
1.9.7. Grid-like shading devices (egg crates, brise soleils)
1.9.8. Frontal shading devices
1.9.9. "Green" shading devices: vegetation as a shading device

Chapter 2. Fundamental Relations Between Greenhouse Features and Climatic Factors

2.1. General considerations
2.1.1. On shape, with regard to solar radiation
2.1.2. On shape, as regards ventilation
2.1.3. Acoustics in greenhouses
2.2. Greenhouses for cold, cool and temperate climates
2.2.1. Additional information about the relations between greenhouse shape and climate
2.2.2. About the slope of the frontal transparent envelope
2.2.3. Relations between the character of daylight and the slope of the transparent enclosures
2.2.4. About east and west enclosures
2.2.5. About roofs
2.2.6. Ventilation openings
2.2.7. Solid thermal masses
2.3. Considerations on greenhouses for cold climates
2.4. Framing the theme of greenhouses for hot climates
2.5. Shadehouses and nethouses

Chapter 3. Fundamental Complements for Solar Greenhouse Design

3.1. On passive heating of greenhouses
3.2. On the role of solar gains
3.3. On the main passive heat transfer strategies in solar greenhouses
3.3.1. On the heat transfer by conduction between greenhouse masses and greenhouse indoor environments

3.3.2. On the heat transfer by conduction between attached greenhouse and building
3.3.3. On the heat transfer by convection between attached greenhouses and buildings
3.3.4. On the types of thermal masses within greenhouses
3.4. On the role of thermal masses for passive greenhouse heating
 3.4.1. On the combination of heat transfer by convection and conduction in attached greenhouses
 3.4.2. On the thermal masses loaded by direct radiation
 3.4.3. On the thermal masses loaded by reflected radiation
 3.4.4. On the thermal masses loaded by convection
3.5. Passive cooling of greenhouses
 3.5.1. The role of thermal masses in the passive cooling of greenhouses
 3.5.2. Thermal mass for thermal inertia
 3.5.3. Thermal mass for coolth storage via "night flushing"
 3.5.4. Natural ventilation
 3.5.5. Wind-driven ventilation
 3.5.6. Criteria for predicting wind flows by means of streamlines
 3.5.7. Stack-effect ventilation
 3.5.8. Mixed – wind-driven and stack-effect – ventilation
3.6. Evaporative cooling
 3.6.1. Direct evaporative cooling
 3.6.2. Indirect evaporative cooling
 3.6.3. Evaporative cooling from still water under still air
 3.6.4. Evaporative cooling with still water and air moving over it
 3.6.5. Evaporative cooling with water in movement in a container or channel, possibly on corrugated surfaces
 3.6.6. Evaporative cooling via water sprinkled by pressure as droplets through nozzles, or falling by gravity
 3.6.7. Evaporative cooling by wetting surfaces and transferring the coolth by convection or conduction
 3.6.8. Downdraught cooling
 3.6.9. Radiative cooling
 3.6.10. Heating and cooling through seasonal storage strategies involving thermal exchange with the ground
 3.6.11. Layout of cooling strategies in bioclimatic charts
3.7. Greenhouse features deriving from use and typology
 3.7.1. Agricultural greenhouses
 3.7.2. Specificities of inhabitable attached greenhouses
 3.7.3. Stand-alone solar greenhouses
 3.7.4. Lean-to, attached solar greenhouses

Chapter 4. Advanced Complements for Solar Greenhouse Design

4.1. Considerations related to shape
    4.1.1. On the symmetry between solar aperture and heat-loss aperture
    4.1.2. On the optimal tilt of front façades
    4.1.3. On the greenhouse "thickness"
    4.1.4. On the greenhouse width
    4.1.5. On the greenhouse height

4.2. Considerations combining shape and construction
    4.2.1. On gable enclosures
    4.2.2. On roofs
    4.2.3. On the greenhouse "knees"
    4.2.4. Rainwater catchment and collection
    4.2.5. Floors
    4.2.6. Additional considerations about the shared wall between the greenhouse and the building
    4.2.7. Stack-effect-driven heat exchange with the ground during daytime

4.3. Ventilative considerations related to shape
    4.3.1. Openings on the greenhouse and the building as regards wind-driven ventilation
    4.3.2. Directionality control for wind-driven ventilation
    4.3.3. Openings in the shared wall with respect to wind-driven ventilation, with the greenhouse front in pressure
    4.3.4. Openings in the shared wall as regards wind-driven ventilation, with the greenhouse front in depression
    4.3.5. Openings in the shared wall as regards wind-driven ventilation, with the wind direction parallel to the fronts
    4.3.6. Combination of stack-effect and wind-driven ventilation using the openings in the shared wall
    4.3.7. Ventilation openings on the greenhouse façades and the roof

4.4. Position of the shading devices
    4.4.1. External shading devices
    4.4.2. Internal shading devices

4.5. Movable thermal insulation

4.6. Microclimates in solar greenhouses
    4.6.1. Cold-sink pits
    4.6.2. Human thermal comfort in solar greenhouses

4.7. Walkways, in growing greenhouses

Conclusion

References

Index

Summary of Volume 3

Introduction

Chapter 1. Prolog – Overview of Types of Transparent Enclosures

1.1. Risks of condensation on transparent enclosures
1.2. Essentials on glass panel enclosures
    1.2.1. Installation of glass panels
1.3. Essentials on synthetic panel enclosures
    1.3.1. Ribbed panels
    1.3.2. Drained flat double panels
    1.3.3. Corrugated panels
    1.3.4. Synthetic films
1.4. Curtain walls
    1.4.1. Detailing at vertical corners
    1.4.2. Detailing the connection between the front wall and the roof
    1.4.3. Detailing the connection between the side walls and the roof
    1.4.4. Detailing the connection between the roof and the building front façade in attached greenhouses
    1.4.5. Detailing the connection between the roof and the back wall (for stand-alone greenhouses)
    1.4.6. Detailing the transition between an opaque part of the roof and a transparent one
    1.4.7. Detailing the vertical joints between greenhouse and building
    1.4.8. Connecting external systems to the mullion-and-transom system
    1.4.9. Pressure caps on roofs
    1.4.10. Bent plates as anchoring devices
    1.4.11. The façade of windows as an alternative to the curtain wall
    1.4.12. The façade of windows at the lintels and at the transoms

1.4.13. Connection between mullion or transoms and windows by means of setbacks in the profiles
1.4.14. Windows and doors
1.5. Glazing with glass panels
    1.5.1. Glass

Chapter 2. Transparent Plastic Enclosures

2.1. Materials for synthetic panels
    2.1.1. Polycarbonate
    2.1.2. Acrylic
    2.1.3. Fiberglass
2.2. Commonalities between flat polycarbonate, acrylic or fiberglass panels
    2.2.1. Working with polycarbonate sheets
    2.2.2. Working with acrylic sheets
    2.2.3. Working with fiberglass panels
    2.2.4. Installation of corrugated polycarbonate, acrylic or fiberglass panels
2.3. Installation of multi-wall polycarbonate or acrylic panels
    2.3.1. Anchorage by direct screwing
    2.3.2. Anchorage by means of channels
    2.3.3. Anchorage by means of profiled connectors embedded in the joints at the edge of the panels
    2.3.4. Anchorage by means of pressure caps in curtain-wall schemes
2.4. Connection of mullions and/or transoms to the transparent panels by means of pressure caps
    2.4.1. Example of simple and low-cost openable wooden window constructions suitable for self-building
2.5. Further considerations related to solar shading
    2.5.1. Shading paints
    2.5.2. External overhangs
    2.5.3. External Venetian blinds
    2.5.4. External horizontal fins
    2.5.5. Internal Venetian blinds
    2.5.6. Internal horizontal fins
    2.5.7. External or internal fixed skylight grids or roof grids
    2.5.8. Detached external canvases/nets
    2.5.9. Attached external canvases/nets/shade cloths/rollers
    2.5.10. External curtains/rollers
    2.5.11. Detached internal canvases
    2.5.12. Attached internal canvases/nets/shade cloths/curtain rollers
2.6. Thermal insulation
    2.6.1. Resistive insulation
    2.6.2. Reflective insulation

2.6.3. Cavities
2.7. Movable insulation
　2.7.1. Movable insulation panels
　2.7.2. Removable insulation panels
　2.7.3. Thermal blankets and curtains
2.8. Solar reflectors
　2.8.1. Diffuse solar reflectors
　2.8.2. Specular solar reflectors
　2.8.3. Fixed solar reflectors
　2.8.4. Movable solar reflectors
2.9. Mechanical systems for operating the openable frames
　2.9.1. Mechanical system for opening rows of openings
　2.9.2. Mechanical systems for opening single windows and skylights
　2.9.3. Actuator control
2.10. Opaque envelopes
2.11. Thermally broken external balconies
2.12. Paints, stains and preservatives

Chapter 3. Film-enclosed Greenhouses

3.1. Characteristics of polyethylene films
3.2. Alternatives to polyethylene films
3.3. Strategies for installing the films
　3.3.1. Fixing strategies entailing rotation
　3.3.2. Fixing strategies not entailing rotation
3.4. Specific challenges in polyethylene-enclosed wooden greenhouses
　3.4.1. Fitting the structure to avoid tearing the films
　3.4.2. Consequences of vapor condensation in film-enclosed greenhouses
3.5. Multiple polyethylene film layouts
　3.5.1. Inflated multiple envelopes
　3.5.2. Inflated ETFE cushions
3.6. Film-enclosed greenhouses for hot climates
　3.6.1. Orientation of hot-climate growing greenhouses
　3.6.2. Solar shading for hot-climate greenhouses
　3.6.3. Natural ventilation for hot-climate greenhouses
　3.6.4. Construction schemes for hot-climate greenhouses, with particular reference to wood
3.7. Framed structural layouts adopting combinations of portal frames
3.8. Bamboo greenhouses

Conclusion

References

Index

Summary of Volume 4

Introduction

Chapter 1. Greenhouse Typologies

    1.1. Stand-alone greenhouse typologies
        1.1.1. At the core of the stand-alone solar greenhouse conception
        1.1.2. Cold frames
        1.1.3. Solar pit greenhouses
        1.1.4. Tall stand-alone greenhouses
        1.1.5. "Non-solar" stand-alone greenhouses
    1.2. Greenhouses serving buildings
        1.2.1. Integrating the direct gain strategy
        1.2.2. Integrating the indirect gain scheme from attached solar greenhouses
        1.2.3. Atria
        1.2.4. Greenhouses as buffer spaces
        1.2.5. The house-in-greenhouse scheme
        1.2.6. Solutions using the ground as primary thermal storage
    1.3. Additional readings

Chapter 2. Calculation Approaches

    2.1. Thermal calculations
        2.1.1. Calculation of the heat transmission through an opaque panel
        2.1.2. Determination of the average temperature of a greenhouse in steady state
        2.1.3. A simplified calculation method of the steady-state temperature in a stand-alone solar greenhouse (experimental)
        2.1.4. Thermal flux through an indirect solar gain system like a solar wall
        2.1.5. Thermal flux through an attached greenhouse
    2.2. Computer simulation as a calculation approach

2.3. Environmental simulation by means of open-source tools
  2.3.1. Basic thermal modeling and simulation criteria
2.4. Structural calculations
  2.4.1. Preliminary structural sizing
  2.4.2. Preliminary structural sizing with open-source simulation tools
  2.4.3. Techniques for exploring the design options on the basis of the simulated performances
  2.4.4. Metamodeling

Chapter 3. Design Studies

3.1. What is still to be said in greenhouse design
3.2. Calimali's greenhouse in Fagnano Olona, Italy. By Greenhouse Design Workshop
3.3. House "D" in Nantes. Xavier Fouquet
3.4. Bioclimatic house in Villeneuve-Tolosane, France – Nycholas Eydoux
3.5. House in Vals, Italy. Studio Albori
3.6. Rehabilitation and extension of the house "AT" in Fagnano Olona. Paolo Carlesso
3.7. Greenhouse from recycled windows at "Casamatta", Gurone, Malnate (Varese), Italy. Marta Robecchi
3.8. House "GdA" in Cairate, Italy. Paolo Carlesso
3.9. A conference greenhouse at Cascina Cuccagna in Milan. Studio Arcò

Conclusion

Afterword

Appendices

Appendix 1: Thermal and Acoustic Properties of Construction Materials

Appendix 2: Strength of Timber According to the Norm EN 338

Appendix 3: Properties of Transparent Materials

References

Index

Other titles from

in

Civil Engineering and Geomechanics

2022

MAZARS Jacky, GRANGE Stéphane
Damage and Cracking of Concrete Structures: From Theory to Practice

2020

SALENÇON Jean
Elastoplastic Modeling

2019

KOTRONIS Panagiotis
Risk Evaluation and Climate Change Adaptation of Civil Engineering Infrastructures and Buildings: Project RI-ADAPTCLIM

LAMBERT David Edward, PASILIAO Crystal L., ERZAR Benjamin, REVIL-BAUDARD Benoit, CAZACU Oana
Dynamic Damage and Fragmentation

PERROT Arnaud
3D Printing of Concrete: State of the Art and Challenges of the Digital Construction Revolution

SALENÇON Jean
Viscoelastic Modeling for Structural Analysis

ZEMBRI-MARY Geneviève
Project Risks: Actions Around Uncertainty in Urban Planning and Infrastructure Development

2018

FROSSARD Etienne
Granular Geomaterials Dissipative Mechanics: Theory and Applications in Civil Engineering

KHALFALLAH Salah
Structural Analysis 1: Statically Determinate Structures
Structural Analysis 2: Statically Indeterminate Structures

VERBRUGGE Jean-Claude, SCHROEDER Christian
Geotechnical Correlations for Soils and Rocks

2017

LAUZIN Xavier
Civil Engineering Structures According to the Eurocodes

PUECH Alain, GARNIER Jacques
Design of Piles Under Cyclic Loading: SOLCYP Recommendations

SELLIER Alain, GRIMAL Étienne, MULTON Stéphane, BOURDAROT Éric
Swelling Concrete in Dams and Hydraulic Structures: DSC 2017

2016

BARRE Francis, BISCH Philippe, CHAUVEL Danièle *et al.*
Control of Cracking in Reinforced Concrete Structures: Research Project CEOS.fr

PIJAUDIER-CABOT Gilles, LA BORDERIE Christian, REESS Thierry, CHEN Wen, MAUREL Olivier, REY-BETHBEDER Franck, DE FERRON Antoine
Electrohydraulic Fracturing of Rocks

TORRENTI Jean-Michel, LA TORRE Francesca
Materials and Infrastructures 1
Materials and Infrastructures 2

2015

AÏT-MOKHTAR Abdelkarim, MILLET Olivier
Structure Design and Degradation Mechanisms in Coastal Environments

MONNET Jacques
In Situ Tests in Geotechnical Engineering

2014

DAÏAN Jean-François
Equilibrium and Transfer in Porous Media – 3-volume series
Equilibrium States – Volume 1
Transfer Laws – Volume 2
Applications, Isothermal Transport, Coupled Transfers – Volume 3

2013

AMZIANE Sofiane, ARNAUD Laurent
Bio-aggregate-based Building Materials: Applications to Hemp Concretes

BONELLI Stéphane
Erosion in Geomechanics Applied to Dams and Levees

CASANDJIAN Charles, CHALLAMEL Noël, LANOS Christophe,
HELLESLAND Jostein
Reinforced Concrete Beams, Columns and Frames: Mechanics and Design

GUÉGUEN Philippe
Seismic Vulnerability of Structures

HELLESLAND Jostein, CHALLAMEL Noël, CASANDJIAN Charles,
LANOS Christophe
Reinforced Concrete Beams, Columns and Frames: Section and Slender Member Analysis

LALOUI Lyesse, DI DONNA Alice
Energy Geostructures: Innovation in Underground Engineering

LEGCHENKO Anatoly
Magnetic Resonance Imaging for Groundwater

2012

BONELLI Stéphane
Erosion of Geomaterials

JACOB Bernard *et al.*
ICWIM6 – Proceedings of the International Conference on Weigh-In-Motion

OLLIVIER Jean-Pierre, TORRENTI Jean-Marc, CARCASSES Myriam
Physical Properties of Concrete and Concrete Constituents

PIJAUDIER-CABOT Gilles, PEREIRA Jean-Michel
Geomechanics in CO_2 Storage Facilities

2011

BAROTH Julien, BREYSSE Denys, SCHOEFS Franck
Construction Reliability: Safety, Variability and Sustainability

CREMONA Christian
Structural Performance: Probability-based Assessment

HICHER Pierre-Yves
Multiscales Geomechanics: From Soil to Engineering Projects

IONESCU Ioan R. *et al.*
Plasticity of Crystalline Materials: from Dislocations to Continuum

LOUKILI Ahmed
Self Compacting Concrete

MOUTON Yves
Organic Materials for Sustainable Construction

NICOT François, LAMBERT Stéphane
Rockfall Engineering

PENSÉ-LHÉRITIER Anne-Marie
Formulation

PIJAUDIER-CABOT Gilles, DUFOUR Frédéric
Damage Mechanics of Cementitious Materials and Structures

RADJAI Farhang, DUBOIS Frédéric
Discrete-element Modeling of Granular Materials

RESPLENDINO Jacques, TOUTLEMONDE François
Designing and Building with UHPFRC

2010

ALSHIBLI A. Khalid
Advances in Computed Tomography for Geomechanics

BUZAUD Eric, IONESCU Ioan R., VOYIADJIS Georges
Materials under Extreme Loadings / Application to Penetration and Impact

LALOUI Lyesse
Mechanics of Unsaturated Geomechanics

NOVA Roberto
Soil Mechanics

SCHREFLER Bernard, DELAGE Pierre
Environmental Geomechanics

TORRENTI Jean-Michel, REYNOUARD Jean-Marie, PIJAUDIER-CABOT Gilles
Mechanical Behavior of Concrete

2009

AURIAULT Jean-Louis, BOUTIN Claude, GEINDREAU Christian
Homogenization of Coupled Phenomena in Heterogenous Media

CAMBOU Bernard, JEAN Michel, RADJAI Fahrang
Micromechanics of Granular Materials

MAZARS Jacky, MILLARD Alain
Dynamic Behavior of Concrete and Seismic Engineering

NICOT François, WAN Richard
Micromechanics of Failure in Granular Geomechanics

2008

BETBEDER-MATIBET Jacques
Seismic Engineering

CAZACU Oana
Multiscale Modeling of Heterogenous Materials

HICHER Pierre-Yves, SHAO Jian-Fu
Soil and Rock Elastoplasticity

JACOB Bernard *et al.*
HVTT 10

JACOB Bernard *et al.*
ICWIM 5

SHAO Jian-Fu, BURLION Nicolas
GeoProc2008

2006

BALAGEAS Daniel, FRITZEN Claus-Peter, GÜEMES Alfredo
Structural Health Monitoring

DESRUES Jacques *et al.*
Advances in X-ray Tomography for Geomaterials

FSTT
Microtunneling and Horizontal Drilling

MOUTON Yves
Organic Materials in Civil Engineering

2005

PIJAUDIER-CABOT Gilles, GÉRARD Bruno, ACKER Paul
Creep Shrinkage and Durability of Concrete and Concrete Structures CONCREEP – 7

Printed and bound by CPI Group (UK) Ltd, Croydon, CR0 4YY
30/08/2023

08106828-0001